More Plastics for Collectors

A Handbook & Price Guide

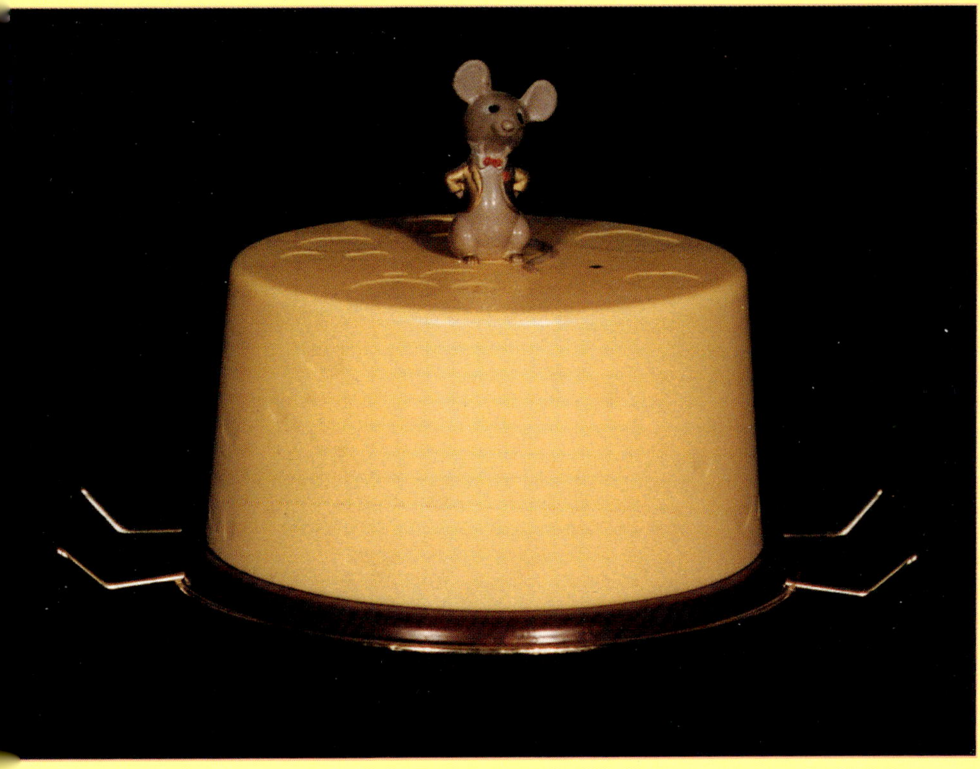

Schiffer Publishing Ltd
7 Lower Valley Road, Atglen, PA 19310

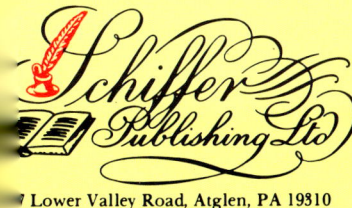

Jan Lindenberger
with Jean Rosenthal

Acknowlegments

I send a very special thank you to Joyce, Kinneth, Jean and Little Kinneth Rosenthal for opening up their home and sharing their wonderful collection for this book. The hours were long and the weather was hot, but with their enthusiasm it was finished with a breeze. Thanks to four kind and generous people for their hospitality. Jean opened up her shop called "Special Somethings" in Lakeport, California, early, and brought many of her "Special Somethings" home to be photographed for this book. Again thanks for all your kindness and help!

I also would like to thank:
Helen Hale, Lakeport, California
Yesteryear, Upper Lake, California
Antique Plaza, Rancho Cordova, California
Cheryl Blackman, Citrus Heights, California
Atomic Antiques & Collectibles, Denver, Colorado
and anyone I may have forgotten

Copyright© 1996 Jan Lindenberger

All rights reserved. No part of this work may be reproduced or used in any form or by any means - graphic, electronic or mechanical, including photocopying or information storage and retrieval systems- without written permission from the copyright holder.

Printed in Hong Kong

ISBN: 0-88740-967-9

Book Design by Audrey L. Whiteside

Library of Congress Cataloging-in-Publication Data

Lindenberger, Jan.
　　More plastics for collectors: a handbook and price guide / Jan Lindenberger with Jean Rosenthal.
　　　p. cm.
　　ISBN 0-88740-967-9 (paper)
　　1. Plastics craft--Collectors and collecting--Catalogs. I. Rosenthal, Jean. II. Title.
　NK8595.L52　1996
　668.4'9--dc20
　　　　　　　　　　　　　　95-42728
　　　　　　　　　　　　　　CIP

Published by Schiffer Publishing, Ltd.
77 Lower Valley Road
Atglen, Pa 19310
Please write for a free catalog.
This book may be purchased from the publisher.
Please include $2.95 postage.
Try your bookstore first.
We are interested in hearing from
authors with book ideas on related topics.

Contents

- Introduction 4
- Plastic in the Kitchen 11
  - Salt & Pepper Shakers 11
  - Bowls & Serving Containers 17
  - Covered Containers 32
  - Butter Dishes & Napkin Holders 46
  - Spoon Rests 52
  - Trivets 53
  - Gadgets 54
  - Dinnerware 67
  - Pitchers 71
  - Placemats & Napkin Rings 75
- Plastics Around the House 76
  - Plastic Flowers & Fruit 76
  - Ashtrays 82
  - Coasters 84
  - Poker Time 86
  - Boudoir Items 89
  - Wall Hangings 97
  - Electrifying Plastic 103
  - Advertising Items 106
  - Banks & Toys 112
  - Miscellaneous Items 117
- Personal Plastics 124
  - Purses 124
  - Bracelets 129
  - Necklaces 135
  - Earrings 137
  - Belts 143
  - Pins 146
  - Hair Clips 149
  - Sunglasses 153

Introduction

This is my second book about collectible plastics. Who would have thought that collecting plastics would be so popular? It is hard to believe that this trend is still going. When one hears you are into plastic collectibles they tend to raise their eyebrows and look at you like you are totally crazy.

It is not, of course, a new phenomenon. Remember when your grandma's old rocking chair caught your eye and you wanted to keep it and she wanted to throw it away. What she saw as old and worn, you saw as beautiful. It is the same with plastics. The world of plastics is interesting and absolutely amazing, and what was once seen as commonplace is now looked at with nostalgia and an eye for its design.

With the today's way of living, plastics have had a tremendous impact. The word plastic, comes from the word "plastikos," meaning able to be molded or formed into many shapes and sizes. If you look around your home you will notice things made of plastic almost everywhere. As the name implies, plastic is often used because it can be shaped into many things other materials cannot. This quality has allowed products and designs to be made that were once impossible. The forms that plastics can take include molding compounds, liquids, adhesives, coatings, solids, foams, laminates, and fibers. Each has its own property and advantage.

Other items were more suitable for plastic than anything else because of the special characteristics it can be given. Things such as jewelry and ink pens use plastic for lightness in weight. Oven proof plastic can withstand high temperatures without melting. The strength of plastics make them great for toys and windows. Radio housings use plastic with special heat resistance qualities. Flooring and counter tops are made in plastic for durability.

On my travels I have found some wonderful plastics. The plastic kitchen items are my favorite, especially the colorful bread keepers, wall pockets, salt and pepper sets, creamers and sugars, clocks, and canisters. It sure livens up a kitchen in a hurry when you see that nostalgic bright red canister set or bread box sitting on the counter. These items are still easy to find and are quite affordable, but the number of collectors of kitchen items is just beginning to increase. In another year the prices will rise too.

The plastic items that have been most collectible over the past few years are Bakelite jewelry, Catalin radios, and the many varieties of celluloid items. Bakelite and Catalin were made from cast phenolics and were used mostly from 1930 to around 1950 when it proved too labor-intensive to be economical. Celluloid, while a wonderful material in many ways, grew unstable and highly inflammable as it aged. It is very difficult for the collector of these plastics to find a bargain, due to their rarity.

Collecting plastics is fun. The colors, shapes, and designs delight the eye and will add to one's personal appearance or to the decor of the home. And in addition it is possible to build a substantial collection without spending a great deal of money...a rarity itself in the world of collecting.

More Plastics for Collectors: A Handbook and Price Guide is designed for the collector. It can be taken to the flea markets, yard sales, and antique shows and used to evaluate a wide variety of plastic items. The information and photographs on these pages will give the collector an overall understanding of plastics. The price guide, though it reflects the prices in the western United States and may differ from other areas, will give the collector a good idea of what an object is worth on the market. (Please note that auction prices will definitely differ from shop prices, and that items will differ in price based on condition and availability.)

I hope you use and enjoy the book, and wish you happy collecting.

Jan Lindenberger,
Colorado Springs, Colorado

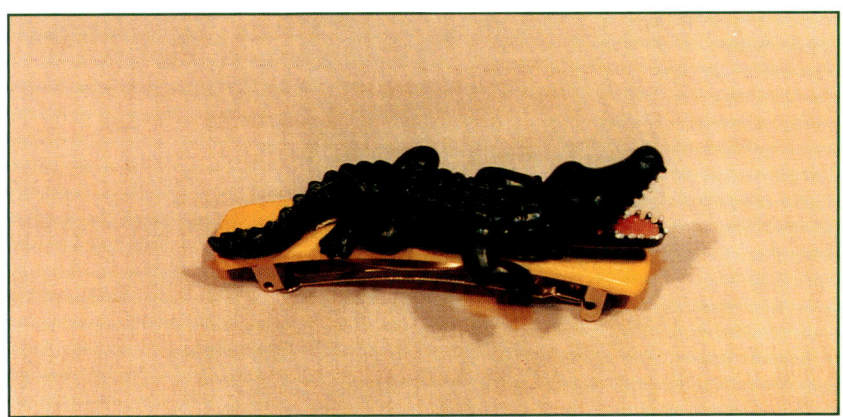

A Brief History of Plastics

The first plastic, cellulose nitrate, was produced in 1862 by Alexander Parkes, an English chemist. This chemical was made from the cells in the walls of plants, cellulose. He found that by adding camphor to the cellulose, it turned into an ivory-like material that could be softened and molded. The product was called Parkesine, after its inventor. Many awards were given him for this product, but it was never in much demand.

Celluloid plastic, a product much like Parkesine, was developed by John Wesley Hyatt in 1868. He began to experiment with collodion, which is a solution of nitrocellulose in ether and alcohol. When exposed to air, it dried into a hard, clear finish. This was used as a finish for wood and metal. Hyatt mixed collodion with camphor and developed celluloid, the first commercially successful plastic. Collars, cuffs, brushes, novelties, dolls, jewelry, toys, billiard balls, vanity sets, shoe horns, are but a few of the items that were made from celluloid in the years that followed.

In 1901, in Germany, Dr. Otto Rohm discovered acrylic plastics. Coal gases, petroleum, air, and water were used to make a clear liquid call methyl methacrylate monomer. This, then, was polymerized into crystal clear, solid plastic. These are a strong and rigid plastics, and are outstanding for outdoor use. Sunlight or the weather has little or no effect on them. Early on, acrylic plastic was used as a bonding agent for safety glass. These plastics are used for outdoor signs, auto light covers, contact lenses, windows, and knobs. Because they are odorless and tasteless, they are also used for juice containers and bowls.

In 1909 Dr. Leo Bakeland made an impressive discovery of a synthetic plastic called phenolic. Also known as Bakelite, it was made by a chemical reaction of formaldehyde and phenol. This plastic was excellent for electrical items like outlets, switches, and auto distributors. It also found use in modern pool balls, telephones, phonograph records, jewelry, toys, and shaver cases.

Cellulose acetate was first developed for commercial use in 1927. Cotton was purified, washed, dried, fluffed, and compressed into bales. These were then treated with acetic acid and acetic anhydride with the aid of a catalyst, such as sulfuric acid. Dyes, lubricants, and plasticizers were then added. When aged and exposed to the ultraviolet rays of the sun, this made a very stable plastic. Lamp shades, lamps, handles, jewelry, toys, combs, dresser sets, and auto parts are molded from this plastic.

Vinyl resin plastic was first made in the United States in 1928 by the Union Carbide and Carbon Corporation under the trade name of Vinylite. This resin was formed by polymerization of vinyl chloride and vinyl acetate. It was used for phonograph records, can linings, electrical insulation, rainwear, air-inflatable items, and food packaging.

Urea-Formaldehyde plastics first appeared on the market in 1929. With the introduction of this resinous compound, it became possible to make plastic in many bright and beautiful colors.

In 1936 more vinyls were developed. These were tough materials which wore well, and were easily embossed and printed. These new vinyls were used for floor covering, rain coats, upholstery, and shower curtains.

Nylon was invented in 1938. It was tough and resilient, with tremendous strength and a low coefficiency of friction. These qualities made it an ideal material for bearings, fishing line, gears, bristles for brushes, and hundreds of other uses for industry, the home, and in clothing.

Other important developments in plastic technology include:

1938 Styrenes. Styrenes are transparent and good at low temperatures. They are commonly used for refrigerator surfaces, food bins, and wall tile.

1939 Aminos. Hard and able to be produced in beautiful colors, aminos were used for buttons, dishes, lamp shades, and colored telephones.

1940 Silicones. Silicone plastics are heat resistant and almost like rubber. They are used for sealants, artificial heart valves, and lubricants.

1942 Polyesters. When combined with glass fibers, polyesters have the strength needed for use in boats, suitcases, sleds, walls, and auto bodies.

1942 Polyethylene. These soft, flexible plastics are not harmed by food or water. They are used for toys, wire insulation, packaging film, and squeeze bottles.

1943 Fluorocarbons. The fact that fluorocarbons are unaffected by chemicals and are nearly frictionless, makes them well-suited for such common uses as non-stick frying pans, valve seats, and bearings.

1947 Epoxies. Because of their adhesion qualities epoxies are widely used in glue, coatings for roads, and floors.

1957 Polypropylene. Its break resistant quality makes polypropylene a good material for industrial moldings, ice cube trays, squeezable bottles, and plastic bags.

1962 Phenoxy. This is an engineering plastic that is rigid and hard. Phenoxy is used for computer parts, bottles, drug containers, and appliance housings.

1962 Polyallomer. Its high flexibility and long life make polyallomer a good material for hinges. It can be colored and processed easily for uses in hinged cases, notebook binders, and containers.

Since 1962 the plastic industry has been in full force, with new types of and uses for plastic being developed almost daily. Today the packaging industry is the largest market for plastics, followed by the electronics industry. Transportation is fourth and furniture fifth. Other large users of plastics include the appliance industry in sixth place and the toy industry in seventh.

This list makes it clear how far reaching and how great the impact of plastic is in our lives.

How Plastics Are Made

Plastics come from natural resources that include petroleum, natural gas, coal, water wood, salt, air, and limestone. These are broken down into carbon, hydrogen, oxygen, and nitrogen atoms, with carbon being the basic component of plastics. The way these elements are combined and treated with heat determines the type of plastic made.

There are two basic ways of creating plastics. In the traditional manner, the raw materials of plastic are changed by the use of heat, pressure, and a catalyst (a substance that causes chemicals to react to each other, but does not itself become part of the final result). The latest technologies avoid the use of heat, pressure, and catalysts by shooting powerful doses of radiation through the raw materials. The results are the same but better control is achieved with this method because the plastic stops changing as soon as the radiation is switched off.

The four principal types of organic plastics are: (1) synthetic resins; (2) natural resins; (3) cellulose derivatives; and (4) protein substances. A brief description of each of these groups will acquaint the reader to the characteristics of each type.

Synthetic resin plastics--When combined with suitable fillers, these plastics are molded into lightweight products with excellent strength and dimensional stability. They also have resistance to the elements which cause deterioration such as moisture and sunlight. Products are rapidly manufactured in large quantities of accurately sized parts by the application of heat and pressure to the material placed in suitable molds. Casting resin and laminated resinous products are made into sheets, rods, or tubes. Machine operations cut blanks from these for the finished product. Some of the cheap raw materials used in the production of resin plastic include formaldehyde, phthalic anhydride, acetylene, and petroleum. This plastic is commonly known under trade names as Bakelite, Catalin, Beetle, Glyptal, and Vinylite. Resin plastics are used in such things as electrical parts, containers, clothing accessories, including buttons, buckles, jewelry, and miscellaneous novelties.

Natural Resins--The natural resins that are the basis of these plastics are commonly known by such names as shellac, rosin, asphalt, and pitch. They are used in industry for the production of the fusible type of molded product as distinguished from the infusible articles formed by some of the synthetic resins. Hot-molding compositions are prepared by mixing shellac, resin, and asphalts with suitable fillers. Compositions containing mainly shellac as the binder are used in insulators for high voltage equipment, telephone parts, and phonograph records.

Cellulose Derivatives--This organic plastic is probably the most used and most best known of any plastic. Celluloid plastics are used for toys, toilet articles, pen and pencil barrels, camera film, safety glass, windows, and lacquers. The raw material cellulose, is found in fairly pure, fibrous condition as ordinary cotton or pulped wood. Treatment with chemicals converts cellulose into compounds which can be formed into desired shapes. Cellulose plastics excel in toughness and flexibility, conduct heat slowly, and can be made tasteless, odorless, and transparent.

Protein substances--These plastics are named according to the protein source of the material; for example, casein plastic from skimmed milk and soybean meal from soybeans. These protein substances are thoroughly kneaded into a colloidal mass, which is then formed into rods, tubes, and sheets by suitable presses. The formed pieces are then hardened by treatment with formaldehyde. Buttons, buckles, beads, and game pieces are made by this process.

The two main groups that plastics fall into are thermoplastics and thermosets. Thermoplastics can be changed into different shapes by pressing or molding them while they are softened by heat. They will keep the new shape when they are cool but when reheated they will soften again and may be reshaped into something quite different. Thermoplastics are made from ethylene, propylene, oil, benzine, chlorine, and salt. Types of thermoplastics include fluorocarbons, cellulosics, polyvinyl chloride, styrenes, vinyls, polypropylene, polyethylene, nylon, and acrylics. Some examples of products they are used for include handbags, ice cube trays, eye glass frames, bowls, doll parts, hair brushes, and golf tees.

Thermosets may also be softened and molded, but only once. After they have been heated they undergo a chemical change so that, when cool and hard, they keep their shape forever and cannot be remolded. The thermosets made of formaldehyde, nitrogen, coal, air, and cellulose sources such as cotton, wood pulp, and corncobs. They include the epoxies, polyesters, aminos, Bakelite, urea, melamine, and phenolics. Some examples of these are bottles, poker chips, makeup cases, knobs, and buttons.

Most chemical names of plastics are complicated and difficult to pronounce. Sometimes the chemical name of the resin is used and in other cases the trade name is used. Some inventors even used "catchy" names. Because of the great variety of terminology and to avoid considerable confusion, the following list of manufacturers and trade names will help the collector identify products. It is impossible to make a complete list because new materials are developed and named each year.

Plastic in the Kitchen

Salt & Pepper Shakers

Candle stick salt and pepper set. 1950s. $15-20

Mammy and chef plastic black salt and pepper set. 1950s. $45-55

Hat rack salt and pepper set. 1950s. $15-20

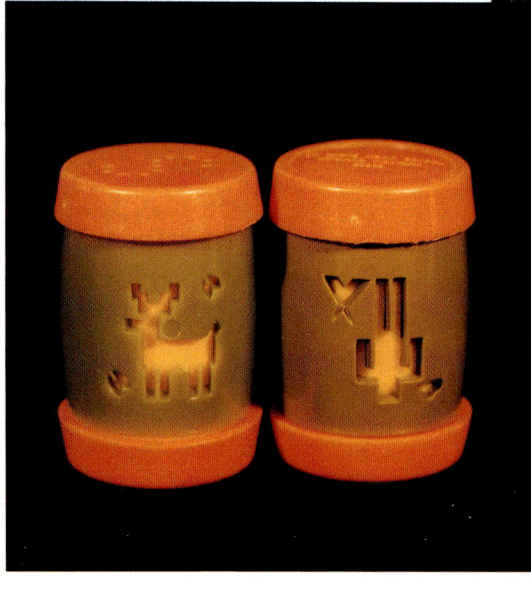

"St. Habre Indian School" salt and pepper set. 1970s. $8-10

Mr. and Mrs. Humpty Dumpty salt and pepper set. 1970s. $15-20

White and red plastic salt and pepper set. 1960s. $4-6

Combination salt and pepper shaker, with partition. Two different colors. "Hong Kong," 1970s. $6-8

Ivory colored cornucopia salt and pepper set. 1960s. $18-20

Mr. and Mrs. Santa mice salt and pepper set. "Hong Kong," 1960s. $12-15

Clear ribbed plastic salt and pepper set with chrome tops. 1970s. $12-15

"Federal House Wares" glass salt and pepper set with plastic tops. 1950s. $8-10

Floral salt and pepper set. 1970s. $5-8

Bowls & Serving Containers

"Texasware" mottled purple and grey bowl. 10". 1960s. $18-22

"Texasware" orange mottled plastic bowl. 8". 1950s. $20-25

"Miramar of Calif." purple mottled plastic bowl. 10". 1960s. $20-25

Orange plastic serving bowl "U.S 1960" $7-10

Fiberglass orange floral bowl. 11". 1960s. $12-15

"Marcrest" Melmac serving bowl. 1950s. $10-12

Fiberglass floral bowl. 1960s. $15-20

"Marcrest" Melmac divided serving bowl with round base. 1950s. $10-12

Plastic divided serving dish. "Japan," 1950s. $20-30

"Marcrest" Melmac divided serving bowl with flat base. 1950s. $10-12

Plastic lacquer shell serving bowl. "Hoveco, Hawaii," 1970s. $20-25

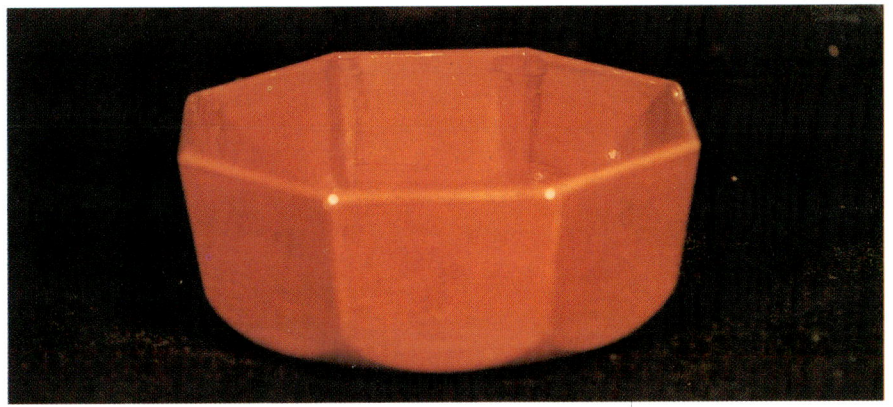

Karian lacquer ware serving bowl. "Japan," 1970s-80s. $8-10

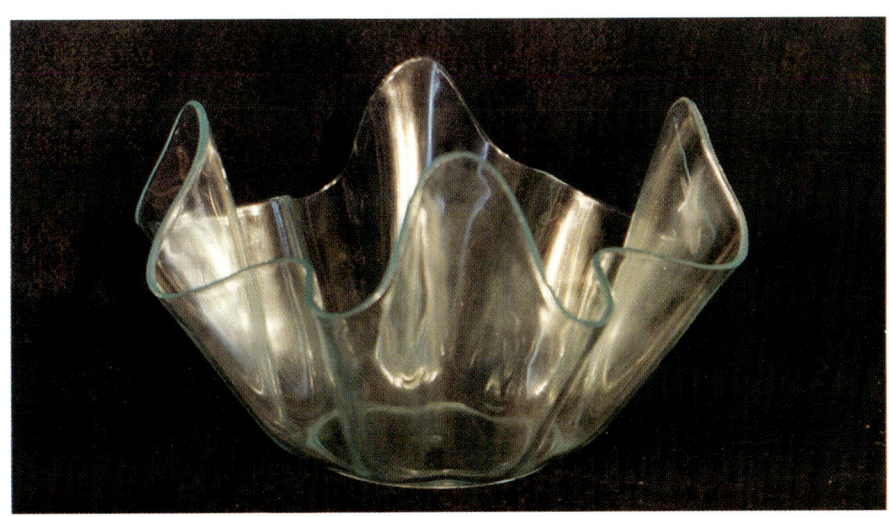

Lucite hankie-shaped bowl. 1960s. $15-20

White laced plastic serving bowl with clear insert. "Regoline" U.S.A., 1960s. $12-15

Pedestal bowl with pierced outer bowl and solid inner bowl. 1960s. $10-15

Lucite nut bowl. 1960s. 6". 1960s. $15-20

Amber and glitter resin nut tray. 12". 1950s-60s. $12-25

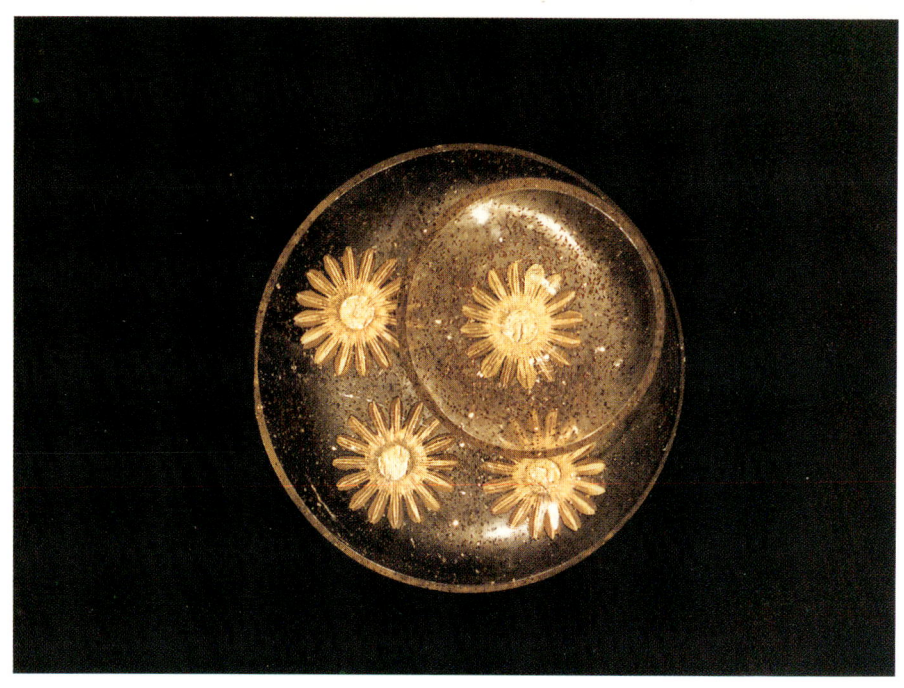

Resin chip and dip tray with glitter finish and gold flowers. 1950s-60s. $20-25

Resin and glitter nut tray with gold leaves. 13". 1950s-60s. $18-20

Three tier resin tidbit tray with metal handle. 14". 1950s-60s. $25-30

Crystal-like plastic tray with metal handle. 12.5". 1950s-60s. $8-10

White diamond cut plastic candy bowl. 1950s. $20-25

Black footed serving tray with turquoise and white design. 1960s-70s. $10-15

Pink flamingo plastic serving tray. 12" x 18". 1990s. $8-12

Fiberglass bowl with sea horse design. 1960s. $12-15

Fiberglass serving tray. 20.5". 1950s-60s. $20-25

Green divided serving tray with mushroom tile in center. 1970s. $8-10

Red divided serving tray with floral tile in center. 1970s. $8-10

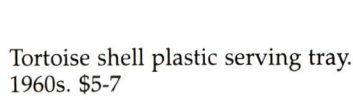

Tortoise shell plastic serving tray. 1960s. $5-7

Resin with bubbles, butterflies and wheat pattern. 15" x 11". 1950s-60s. $18-22

Resin mosaic tile serving tray with wood bottom. 9" x 19". 1950s. $30-35

Resin serving tray with star burst design. 1960s. $10-15

Lace plastic serving tray with floral design in center. 1960s. $15-20

"Texasware" serving tray. 1950s. $10-14

Red fiberglass serving tray with no handles. 14". 1970s. $6-10

"Boltabilt" Bakelite tip tray. 5" x 7". 1960s-70s. $10-14

Red plastic serving tray. 15.25". 1970s. $8-12

Covered Containers

Orange plastic carrot keeper. 9".
1960s. $8-10

Yellow plastic celery keeper. 10".
1950s-60s. $8-10

Set of three cream plastic, garden design canisters. "Sterilite," 1970s. $10-15

Set of three white plastic, garden design canisters. "Sterilite," 1970s. $10-15

White plastic nesting nut dishes with floral motif. 1960s-70s. $12-15

Clear plastic fruit ripening bowl. "California," 1970s. $15-18

Set of three stacking canisters. 1970s. $20-25

Set of four orange plastic canisters. "Tupperware". 1970s. $15-20

Set of three floral canister tins with plastic lids. 1960s. $12-15

Set of four plastic floral canisters. 1970s. $15-20

Set of four plastic green floral canisters. 1970s. $15-20

Set of four glass canisters with plastic lids. "Pyrex". 1960s. $25-35

Set of four red plastic canisters. 1950s. $25-35

Orange "Tupperware" canisters with lids. Set of three. 1970s. $8-10

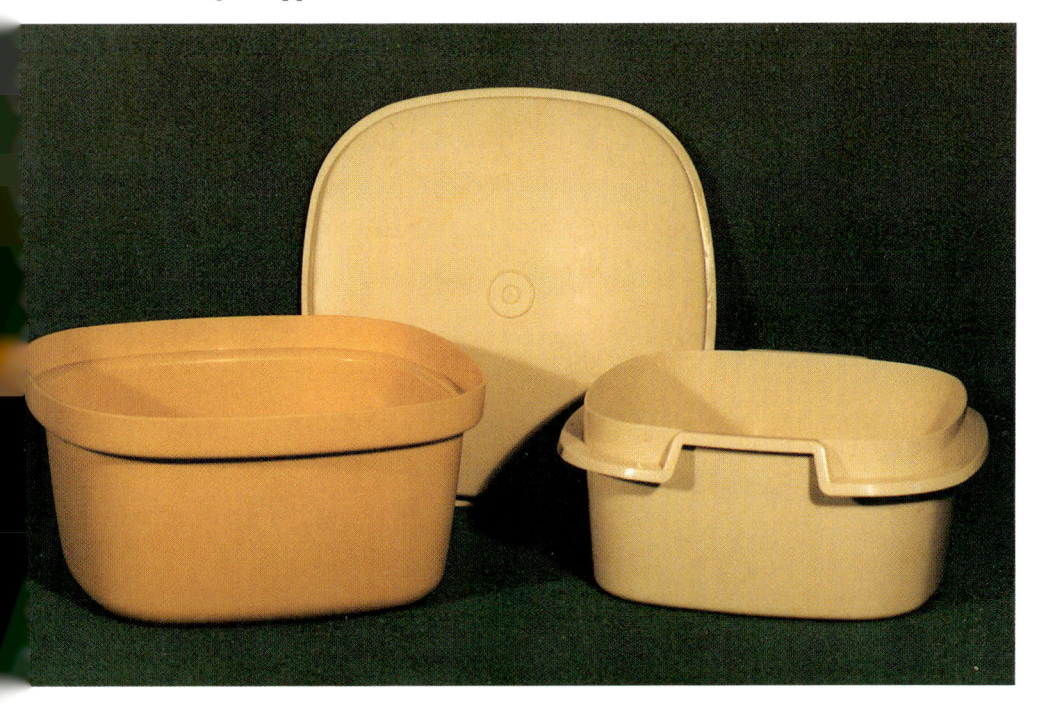

"Tupperware" lettuce keeper with cream color insert. 1970s. $12-15

39

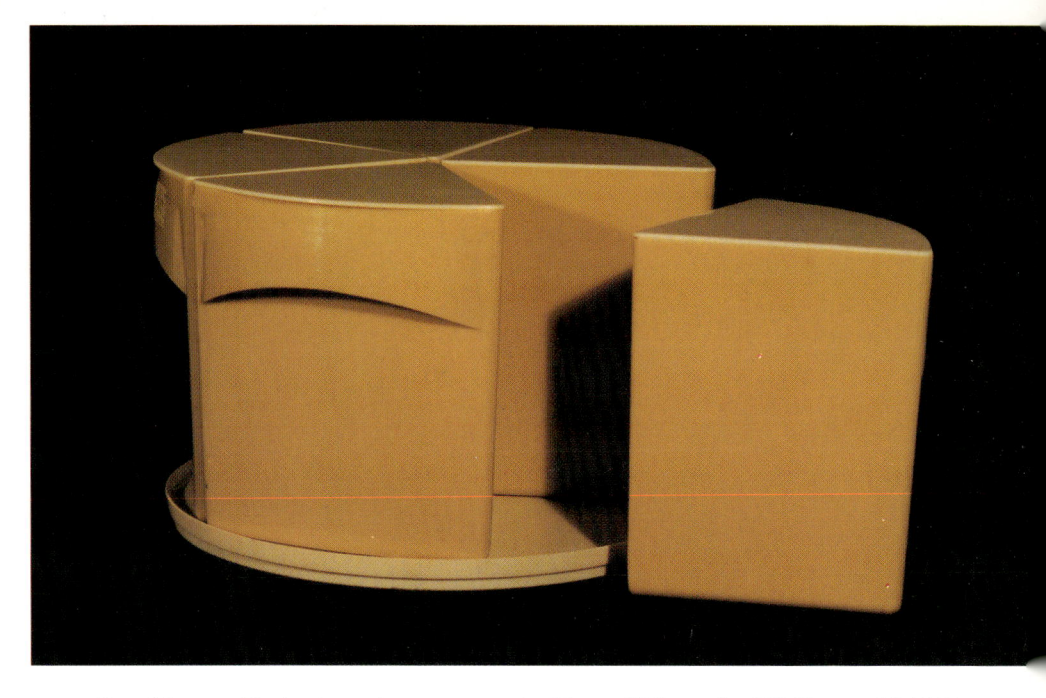

Set of four gold plastic canisters on a swivel base. "Blisscraft of Hollywood," 1960s-70s. $25-35

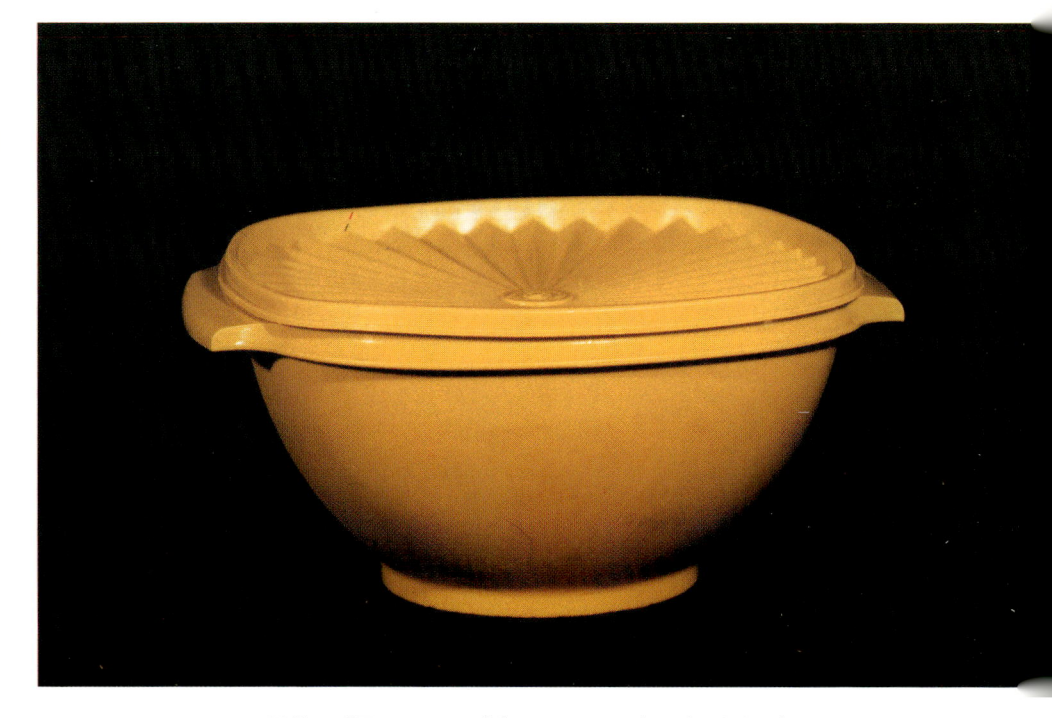

Yellow "Tupperware" large mixing bowl. 1970s. $4-6

Yellow "Tupperware" covered bowls. 1970s. $6-8

"Tupperware" cake keeper. Cream with white top. 1970s. $20-25

Plastic ice bucket, white and black. "Brillium Metal Corp.," 1950s. $35-45

Ice bucket, plastic in wire holder. "Woodpecker Woodware, Japan," 1950s. $12-15

Oriental five piece snack set. Hard plastic. 1970s. $20-25

Plastic cheese keeper with plastic mouse on top. 1960s. $15-20

Pre-war covered swivel cake plate. $25-35

Floral plastic jelly dish. "Emsa" W. Germany, 1960s. $6-8

Clear plastic cream and sugar set. "Blisscraft of Hollywood," 1950s. $15-20 set

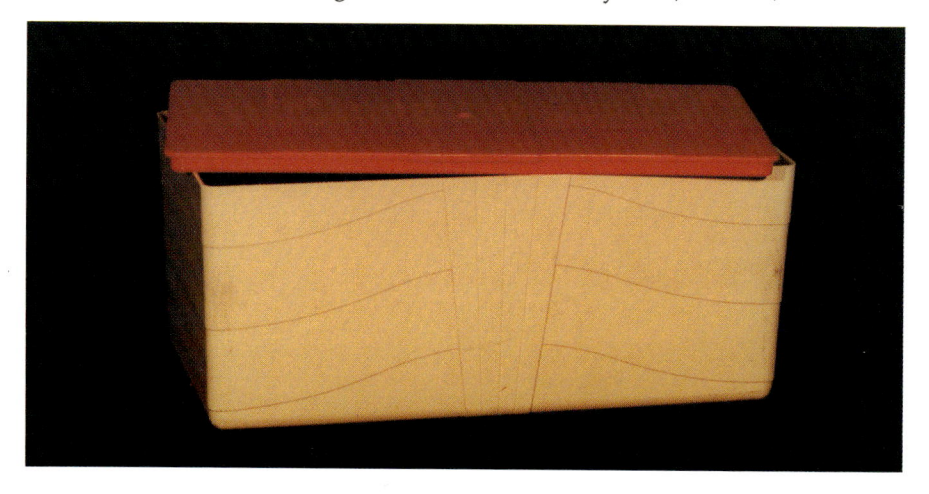

Storage box, cream with red plastic lid. 1970s. $8-10

Aluminum Jell-o mold with plastic serving plate. 1960s-70s. $15-18

Butter Dishes & Napkin Holders

Capital design resin butter dish with imbedded dried flowers. 1960s. $10-12

Butter dish. "Max Klein Inc.," 1970s. $6-10

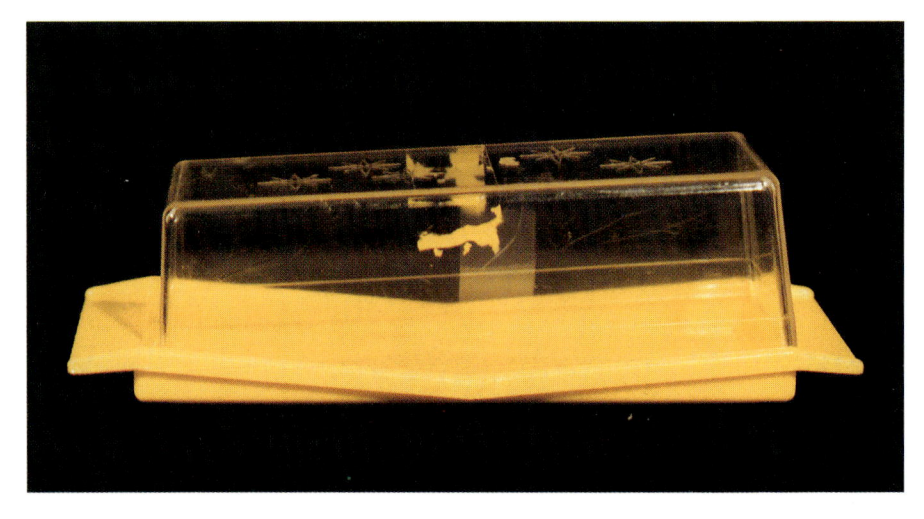

Butter dish. "Made in U.S.A.," 1950s. $7-10

Gold plastic butter dish with lid. 1970s. $5-7

Melmac butter dish with lid, "K La Moune" designer. 1960s. $8-10

"Lusterware" yellow butter dish with lid. 1960s. $6-8

Light weight expandable plastic napkin holder. "Israel," 1970s. $5-7

Resin cast napkin holder with imbedded dried flowers. 1960s. $12-15

Pineapple napkin holder, resin. 1960s-70s. $15-20

Amber plastic napkin holder. "Fresco," 1970s. $7-10

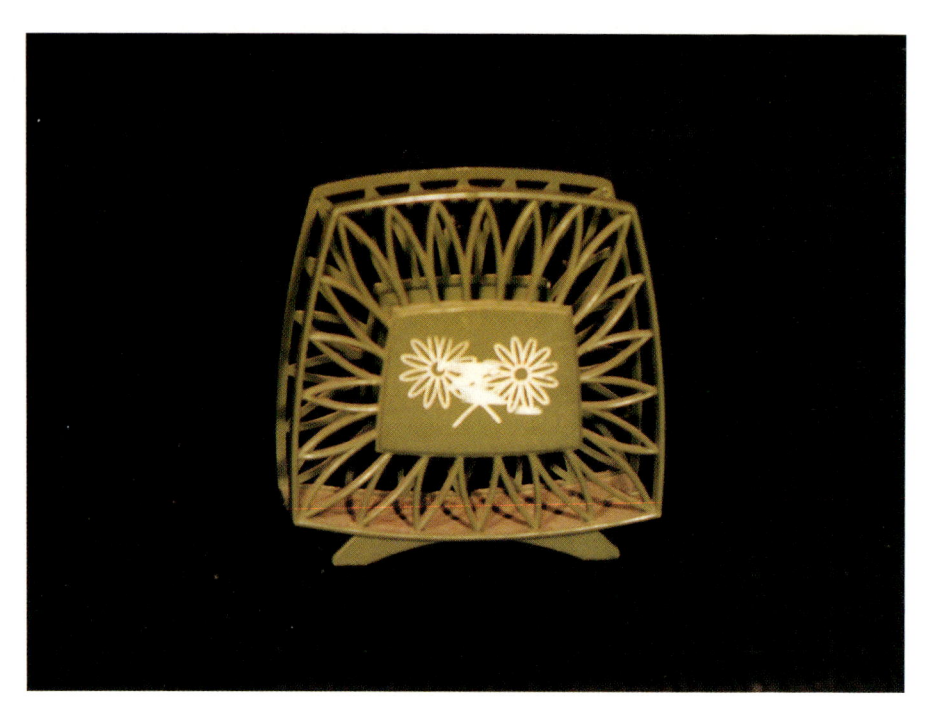

Green plastic napkin holder with daisy design. 1970s. $6-8

Green and yellow flower napkin holder. 1960s. $12-15

Napkin holder decorated with strawberries. 1960s. $6-10

Floral napkin holder. 1970s. $6-8

Spoon Rests

Daisy spoon rest. Resin, 1960s-70s. $5-7

Psychedelic design resin spoon rest. 1960s. $6-9

Resin spoon rest with embedded seeds and dried flowers. 1960s-70s. $8-10

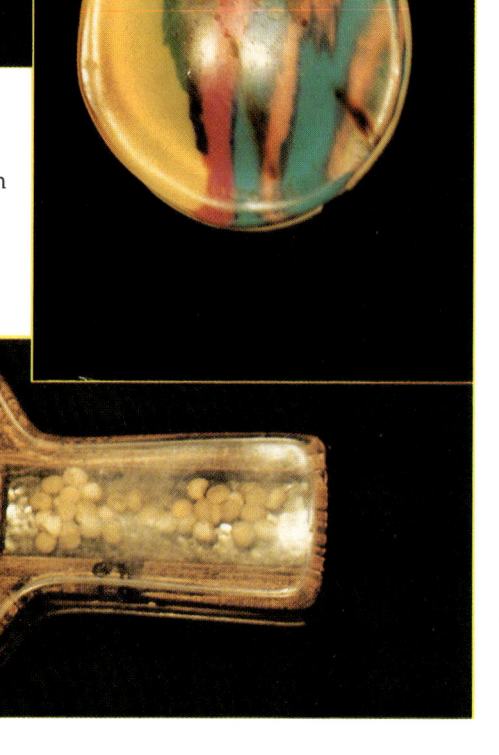

Trivets

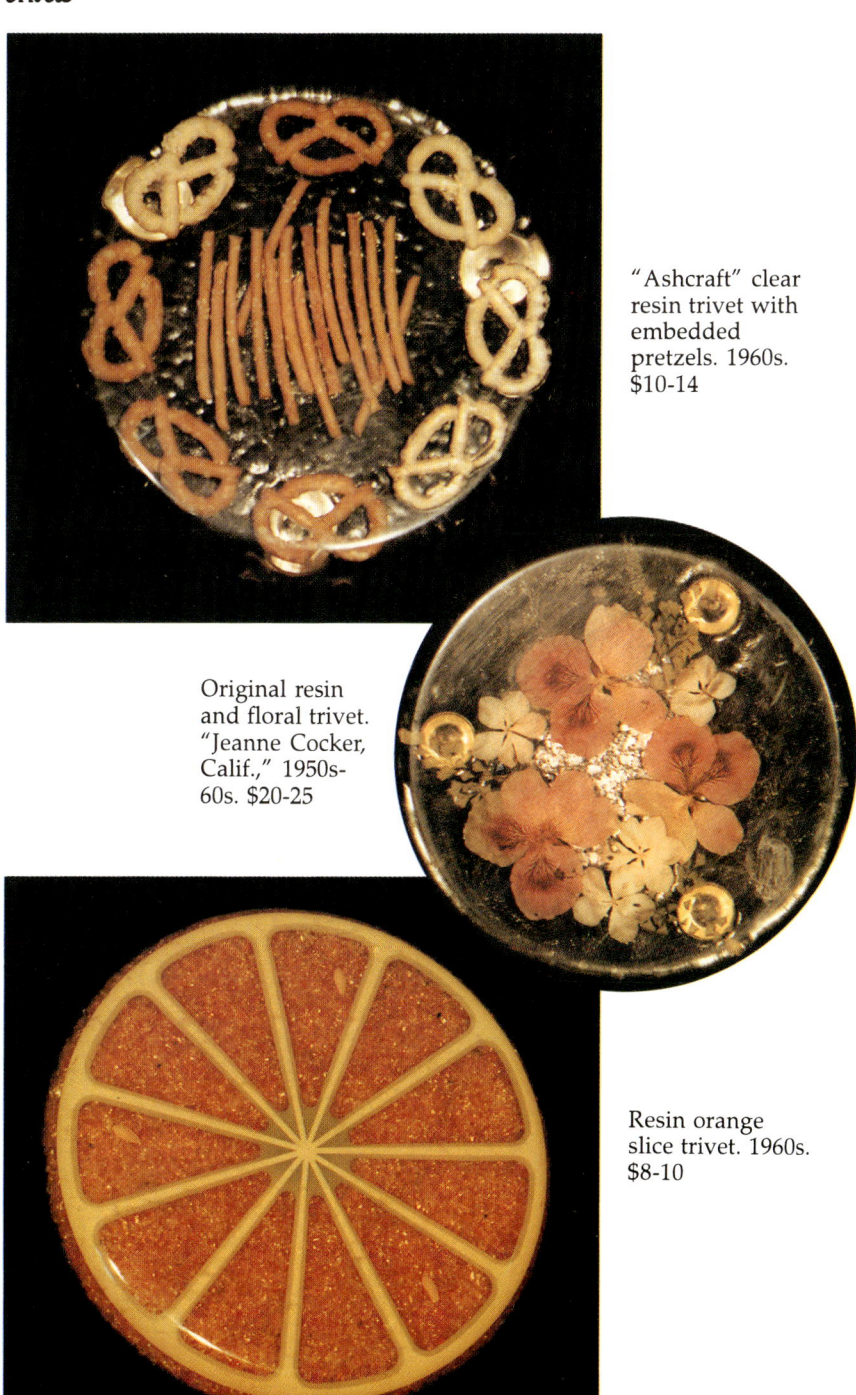

"Ashcraft" clear resin trivet with embedded pretzels. 1960s. $10-14

Original resin and floral trivet. "Jeanne Cocker, Calif.," 1950s-60s. $20-25

Resin orange slice trivet. 1960s. $8-10

Gadgets

Set of six plastic kitchen utensils. "Ecko," 1970s. $35-40 set

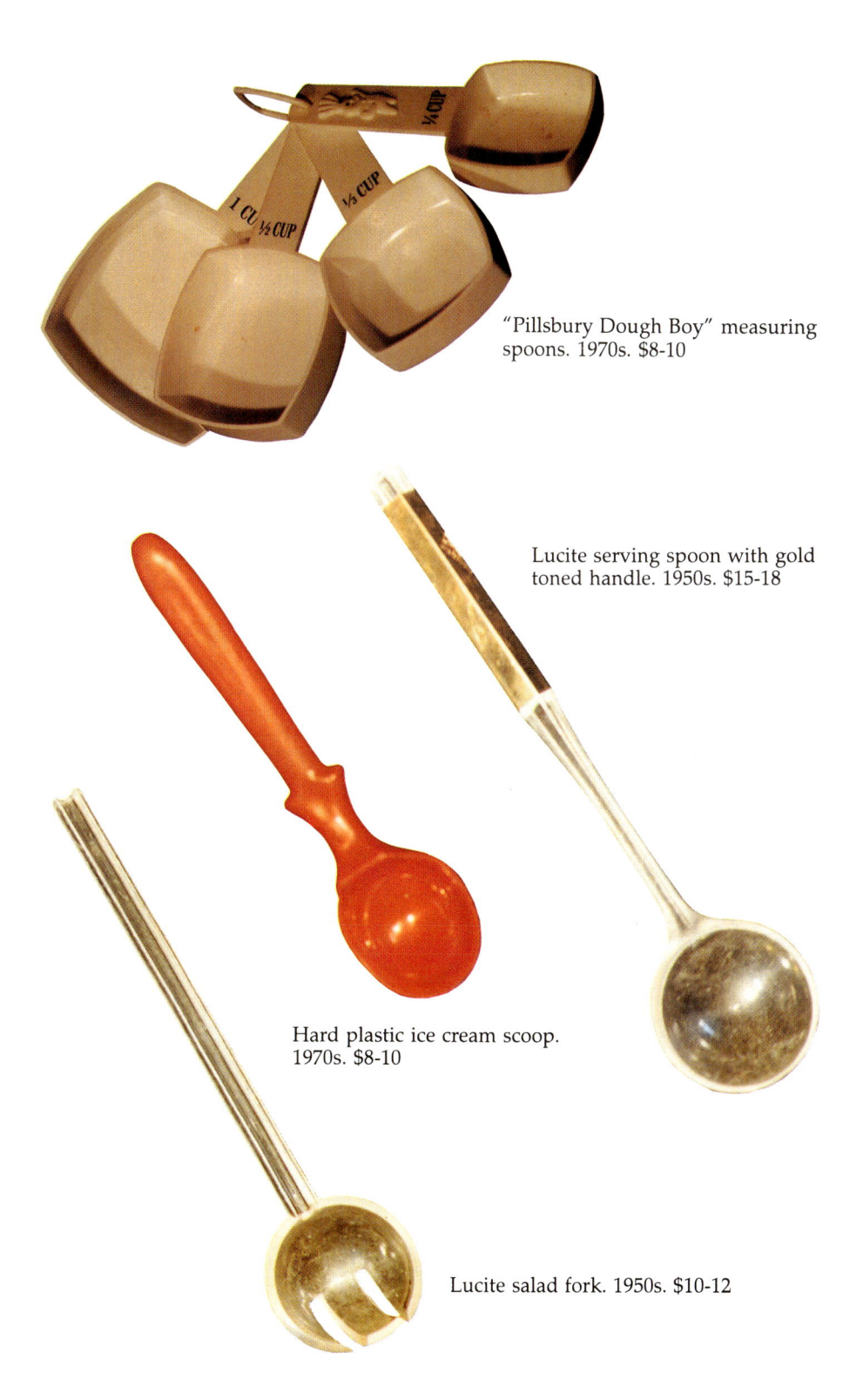

"Pillsbury Dough Boy" measuring spoons. 1970s. $8-10

Lucite serving spoon with gold toned handle. 1950s. $15-18

Hard plastic ice cream scoop. 1970s. $8-10

Lucite salad fork. 1950s. $10-12

55

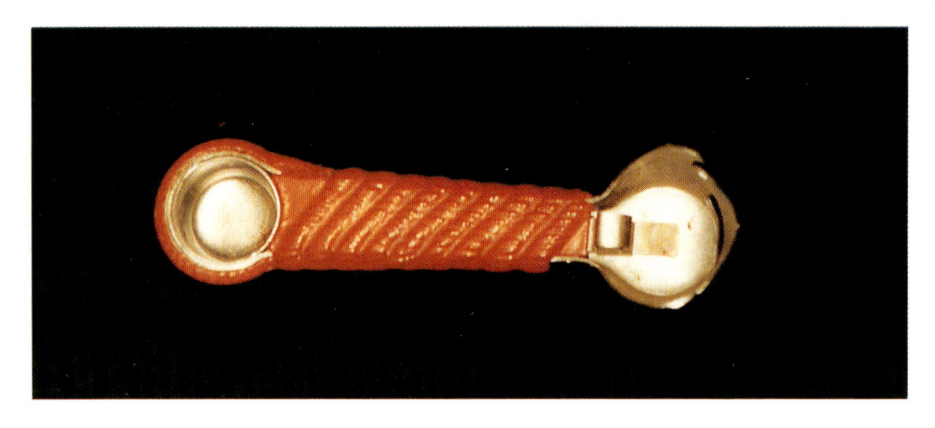

Red plastic and chrome bottle opener. 1950s. $8-12

"Safetyware" red plastic measuring scoop. 1960s. $5-7

Plastic spatula. 1970s. $6-8

Plastic ice cream scoop. 1970s. $5-7

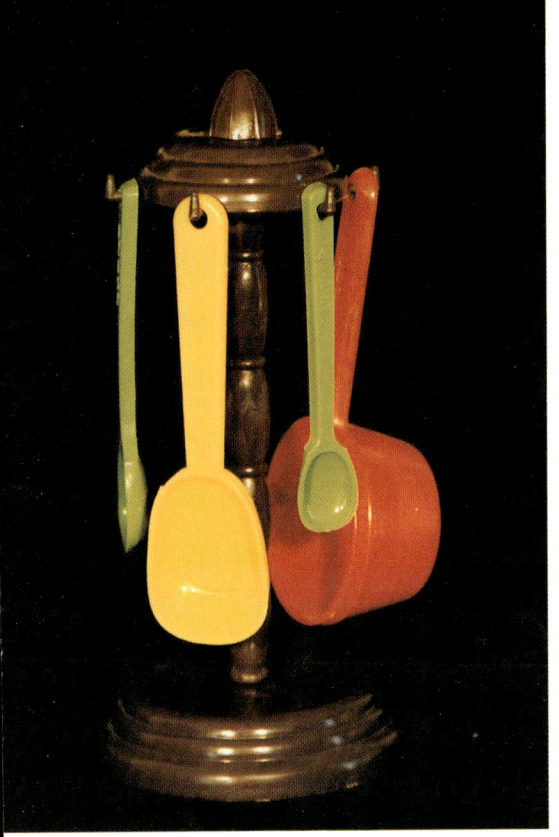

Plastic measuring spoons on wooden rack. "Hong Kong," 1970s. $6-10

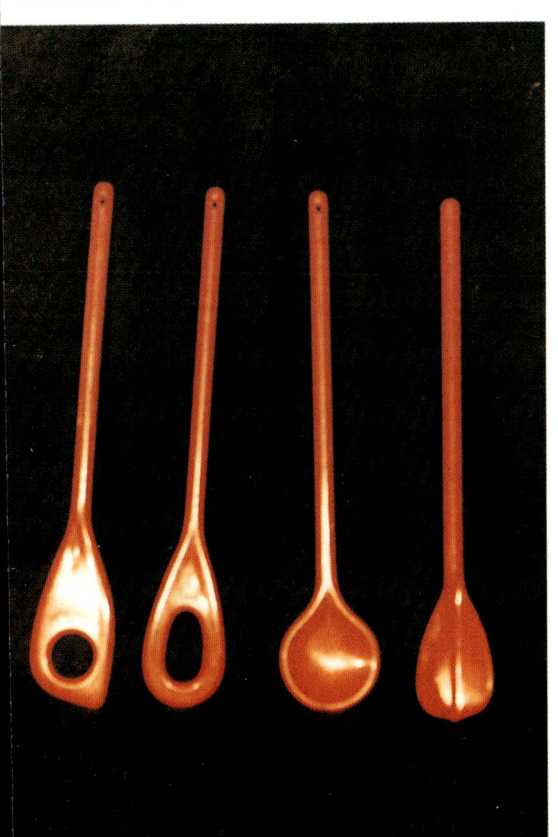

Pear-shaped spoon rest. 1950s-60s. $5-8

Set of four plastic spoons. 1960s. $20-25 set

57

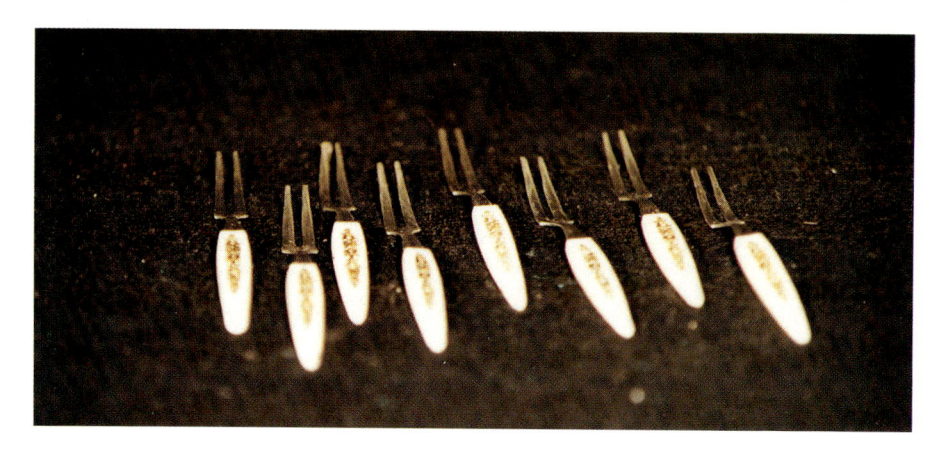

Set of eight snack forks. "Japan," 1960s. $6-8

Set of seven drink stirrers. 1970s. $6-8

Carving set in holder. Ivory colored plastic handles and holder. 1970s. $50-60

"Royal Pacific" plastic measuring spoons with wooden holder. 1960s. $7-10

"Sprout Ease" sprouting tube, plastic. 1960s-70s. $5-8

Donut cutter with yellow plastic handle. 1950s-60s. $5-8

"Hutzler" plastic hand grater. 1970s. $5-8

Chop-o-matic food chopper, plastic and metal. 1950s. $8-10

Yellow "Tupperware" pickle canister. 1970s. $4-6

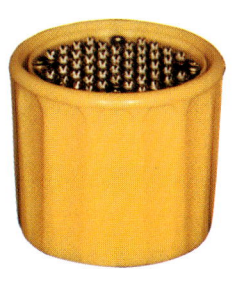

Yellow plastic cheese grater. 1960s-70s. $4-6

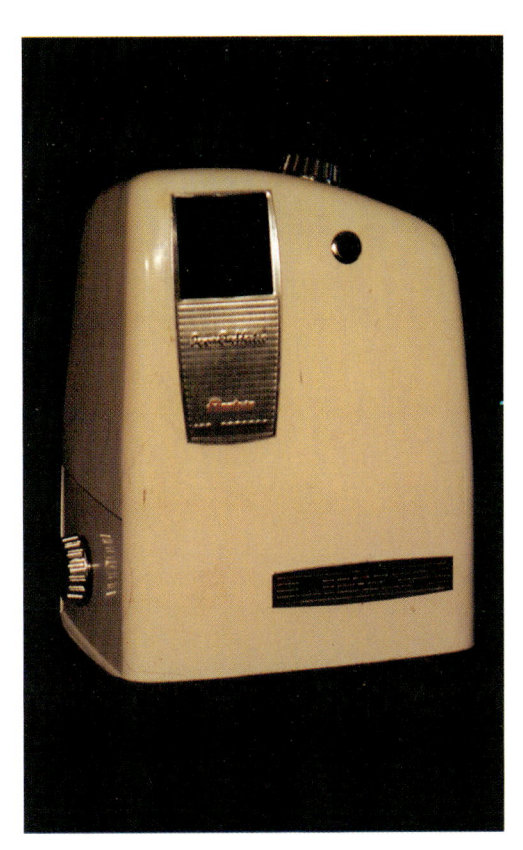

"Rival" electric ice crusher. 1970s. $25-30

Plastic mixer and pitcher. "Made in China," 1970s. $20-30

61

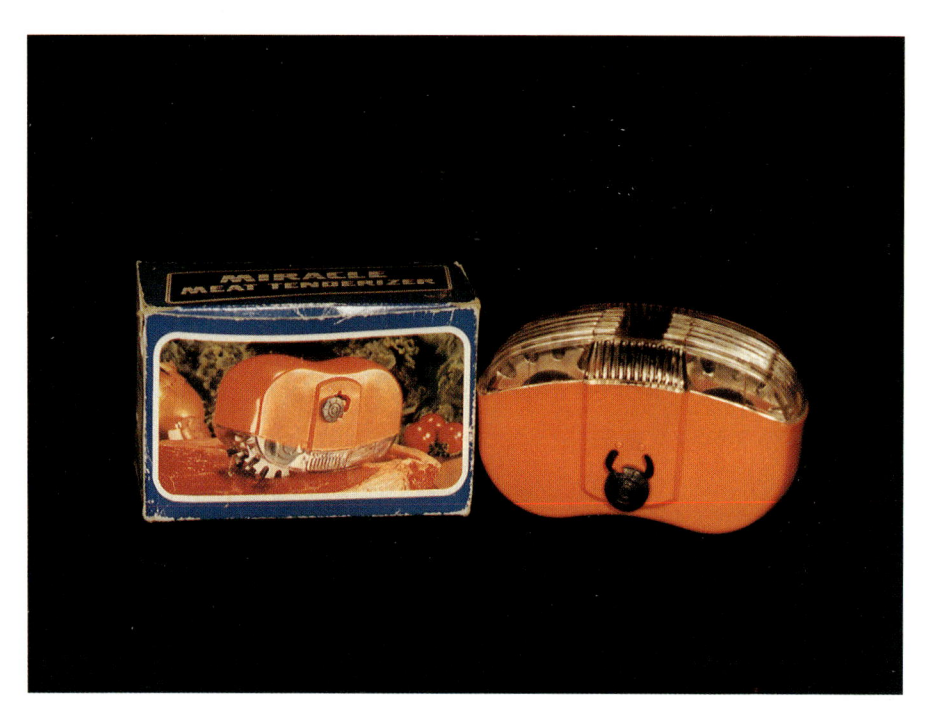

Plastic and metal "Miracle Meat Tenderizer." 1970s. $8-10

"Pro System," battery operated plastic juicer. 1970s. $15-20

Electric "Dorby" mixer with Bakelite handle. 1940s-50s. $30-40

Cheese grater with plastic lid and handle. 1960s. $15-18

Syrup jar with plastic pourer and handle. 1950s. $12-18

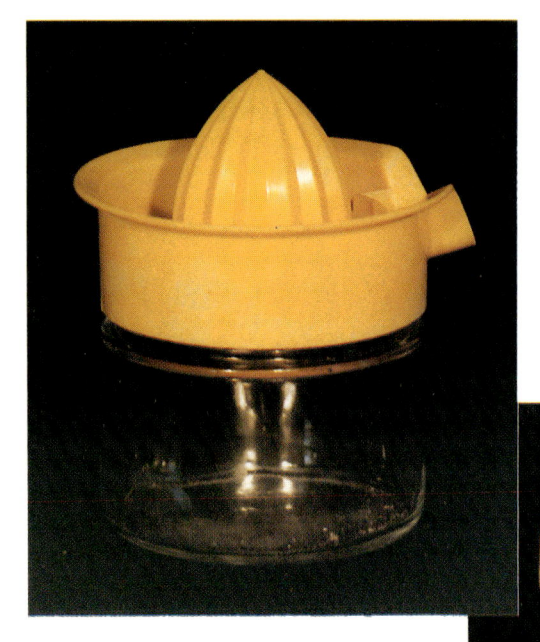

Plastic juicer with glass bottom. 1970s. $8-10

Syrup jar with plastic handle and pourer. "Federal Tool," 1950s. $12-15

Individual coffee carafes, glass with plastic handles and caps. 1960s. $8-12

"Sunkist" yellow juicer with cylindrical bottom. 1960s. $15-20

Juicer marked "Navco, British Hong Kong." 1960s. $8-10

Red plastic juicer with wide rim. "Lusterware". 1950s. $8-12

Pre-war swirl pattern hard plastic juicer. $10-15

Gold plastic juicer with slight rim. 1960s. $5-8

Orange plastic juicer. 1950s. $6-9

Gold plastic citrus juicer. 1960s. $6-8

Red juicer, no rim. "Westland Plastics Inc.," 1950s. $7-10

Dinnerware

Dinnerware service for four. Peach with white flecks. 1950s. $60-75

Creamer, sugar, gravy boat and butter dish. Peach with white flecks. 1950s. $30-40 set

Set of six designer plastic dinner plates. "Heller design by Massimo Vignelli," 1960s-70s. $40-45 set

"Tupperware" Christmas drinking cups with coasters. 1970s. $5-7 each

"Tupperware" cups and coasters/lids. 1970s. $3-4 each

Set of four bright pink ice cream dishes with spoons. 1980s. $15-20 set

Plastic dinnerware, service for four. "Made in U.S.A." 1970s. $35-40

"Melmac" plastic dinner set for four. 1950s-60s. $30-40

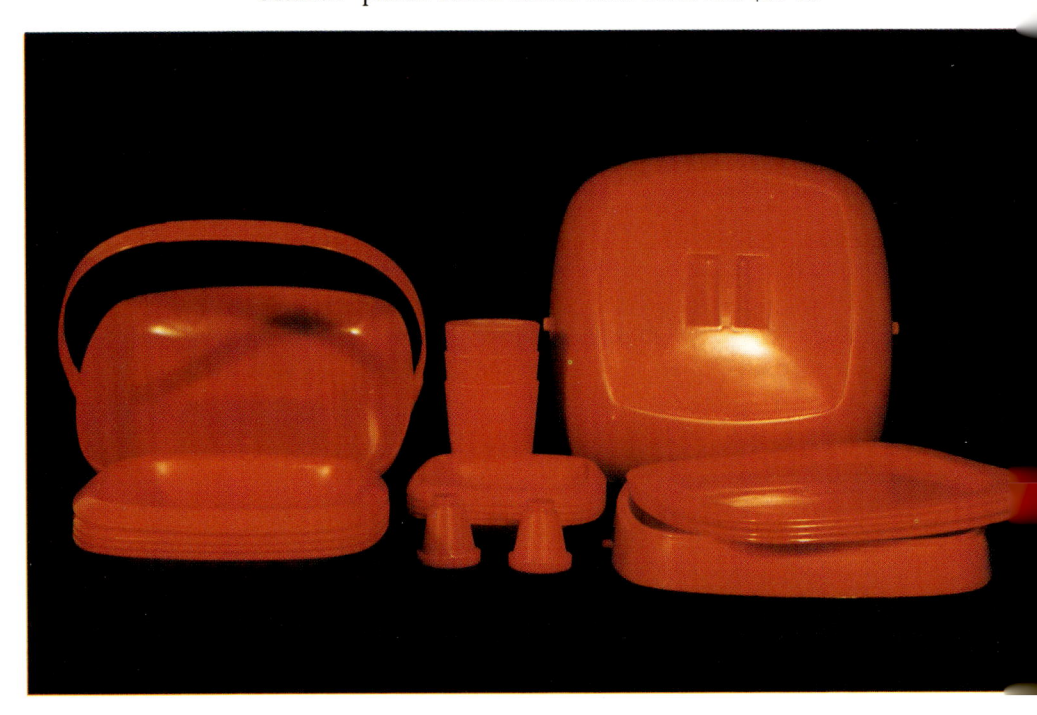

Red plastic picnic set. "Hong Kong," 1970s. $20-25

Pitchers

Hard plastic insulated water pitcher with pink flamingo design. Newer. $20-25

"Aladdin" water pitcher with Pennsylvania Dutch flowers and birds design. 1950s-60s. $15-20

Pink plastic water jar and cup. 1960s. $10-15

"West Bend" plastic Budweiser pitcher. 1980s. $15-20

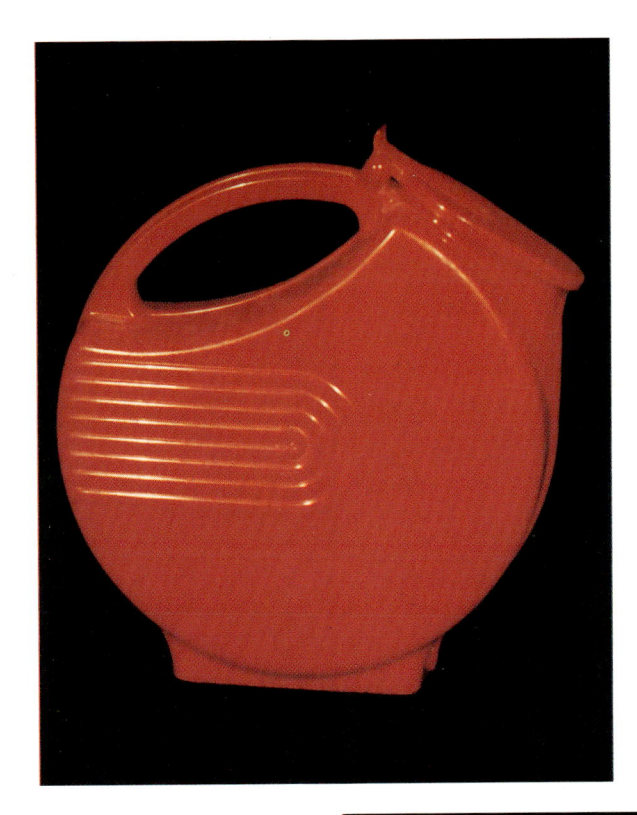

Red plastic water pitcher. "Burrite," 1950s. $10-15

Yellow and cream round water pitcher. Marked "#365". 1950s. $12-15

73

Hard plastic pitcher and two glasses. "Thermoware" by David Douglas. 1970s. $30-40

"Geni" thermos-type pitcher with four cups and four mugs. 1970s. $40-50

Placemats & Napkin Rings

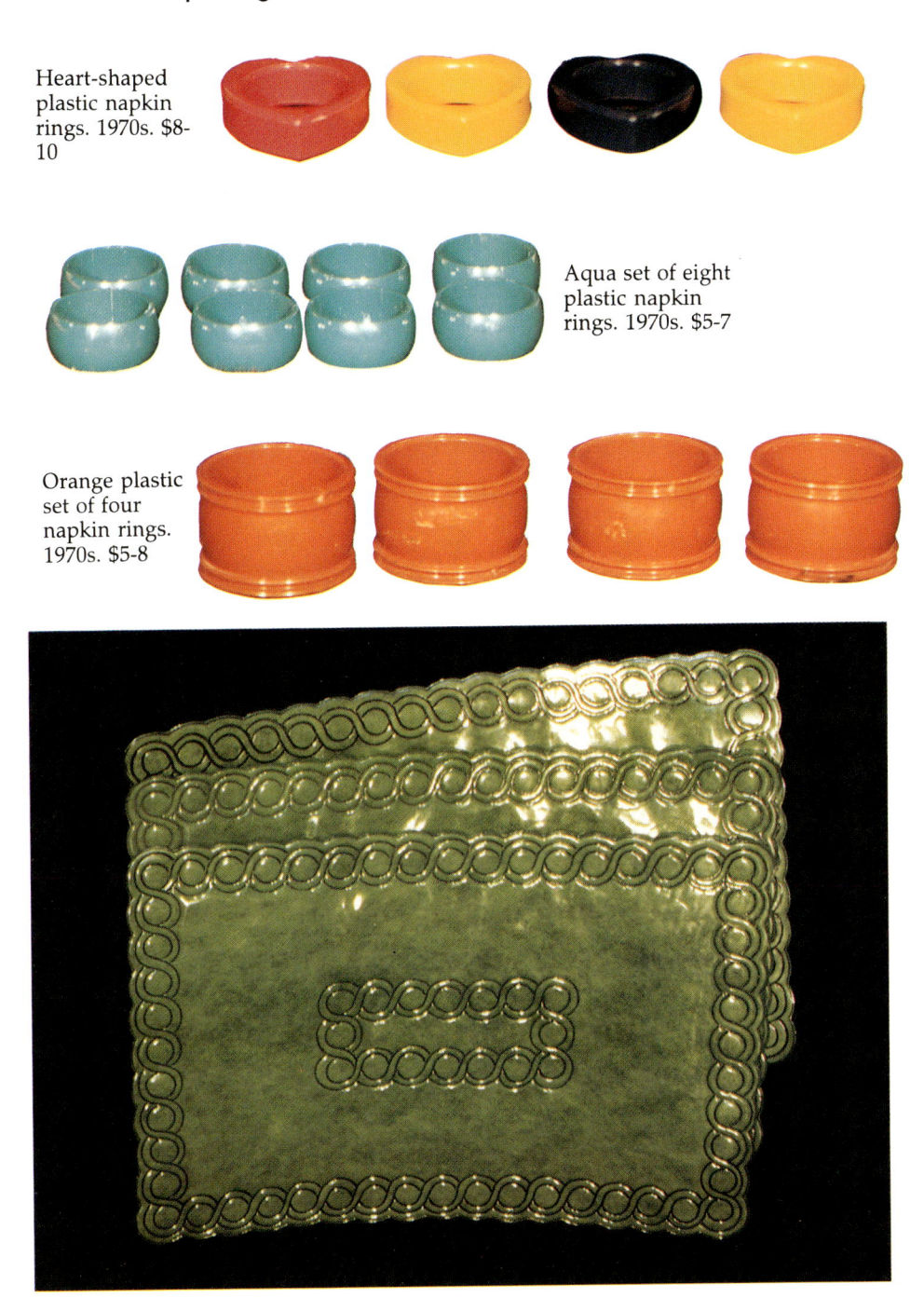

Heart-shaped plastic napkin rings. 1970s. $8-10

Aqua set of eight plastic napkin rings. 1970s. $5-7

Orange plastic set of four napkin rings. 1970s. $5-8

Set of three embossed pattern placemats. 1950s. $10-15 set

75

Plastics Around the House

Plastic Flowers & Fruit

Lucite floral arrangement. 1960s. $25-35

"L.Corelli" plastic flower. 1967. $40-45

Amber and green plastic flower. "R. Iad Co. Ltd.," 1960s. $35-40

Gold and purple soft plastic grapes and leaves. "Hong Kong," 6". 1960s. $6-8

Colorful plastic strawberries with plastic leaves. 1960s. 4.5". $8-10

Hard plastic fruit. 4"-5". 1960s-70s. $20-25 set

Decorative green beans. 4" long. 1960s. $8-10

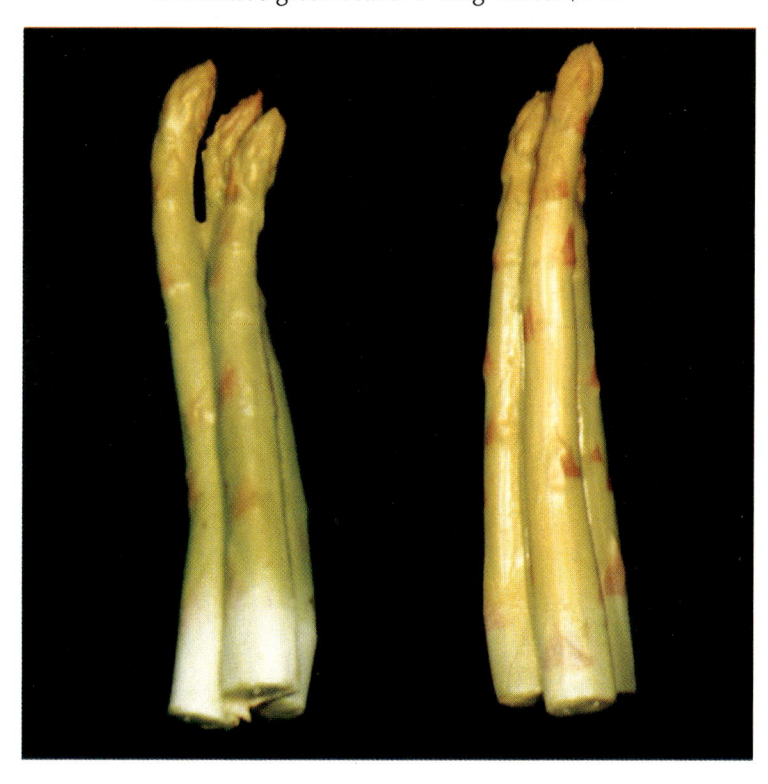

Decorative asparagus. 8" long. 1960s. $8-10

Amber resin decorative grapes cluster. 6". 1960s. $18-25

Two bunches of green florescent decorative grapes. 1960s. 8". $15-20 pair

Red grapes. 8" long. 1960s. $10-14

Orange soft plastic decorative grapes. 1960s. 10". $10-14

Soft plastic yellow grapes with plastic leaves. 1960s. 10". $12-15

Amber and orange resin grape cluster. 20". 1960s. $25-35

Gold pearlized resin grapes. 12". 1960s. $30-35

Red soft plastic grapes with leaves. 1960s. 12". $10-15

Resin ashtray inlaid with gold flecks and sea shells. 1960s. $15-18

Red plastic ash tray. "Made in U.S.A," 1970s. $5-7

Plastic ash tray. "Willert Home Products," 1970s. $3-5

Set of four Bakelite ash trays. 1960s. $15-20

Set of five Bakelite ash trays. 1960s. $20-25 set

Bakelite and metal deco-style cigarette container with roll top lid. 1930s. $45-55

Coasters

Fashion Color Plastic Coaster. "Hallmark," 1970s. $10-15 set

Tortoise shell plastic cork lined coasters. "U.S.A.," 1970s. $8-10

Set of four cork-lined coasters. Fish shaped. 1960s. $6-8

Set of eight floral design fiberglass coasters. 1960s. $8-10

85

Poker Time

Battery operated Card Shuffler. "Jobar International Co.," 1987. $15-20

Poker chip holder. "Tri State Plastic," 1960s. $15-20

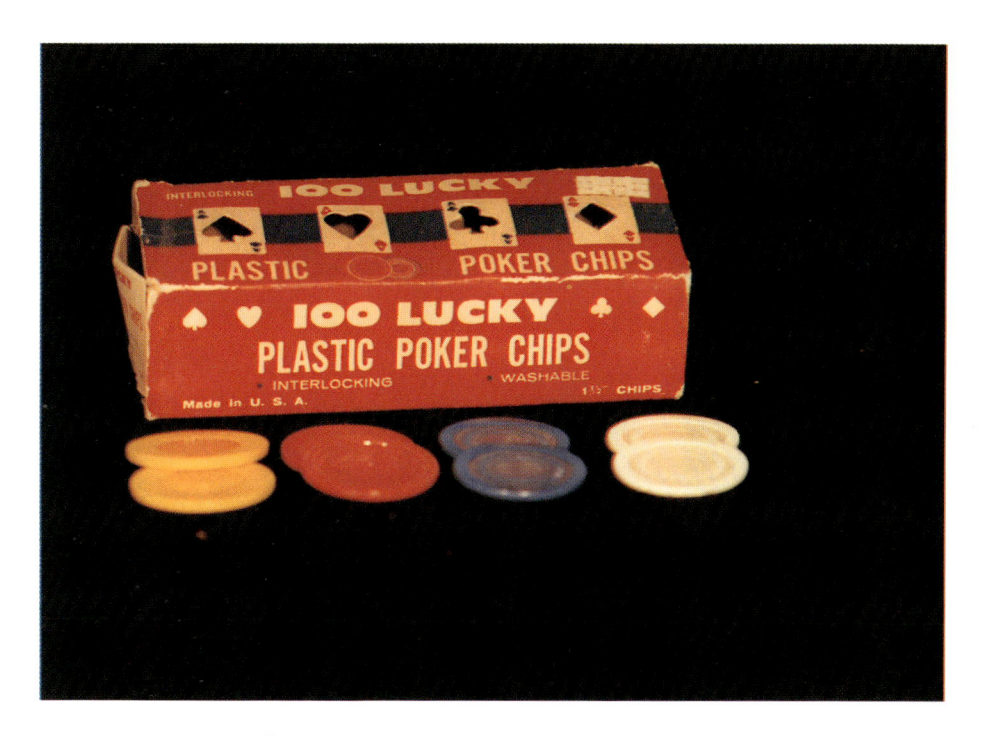

Plastic poker chips in the box. 1960s. $15-20

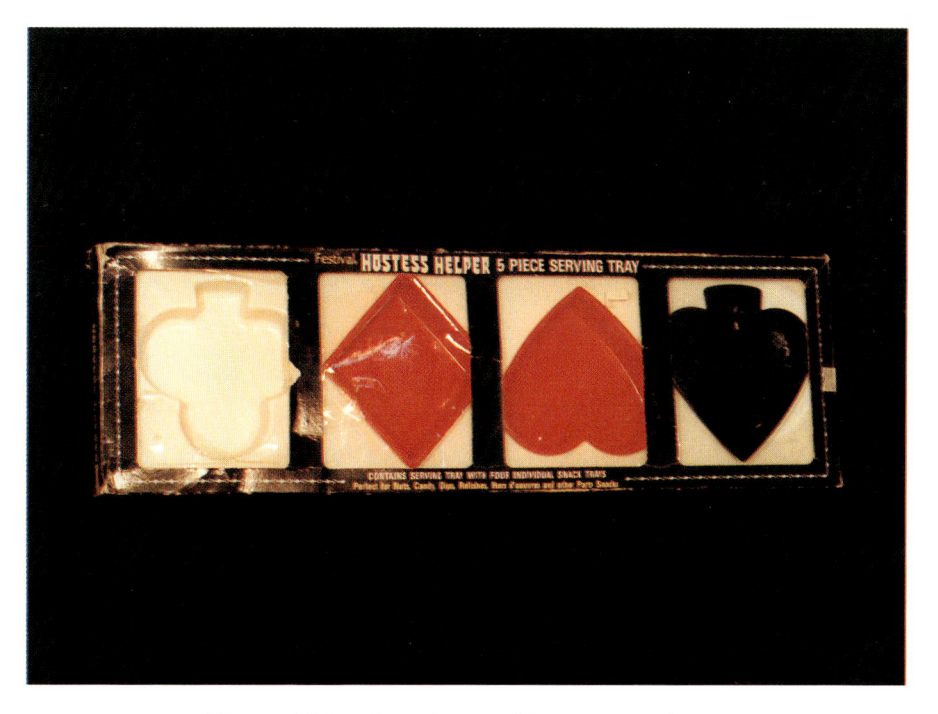

"Hostess Helper" serving set. Plastic, 1970s. $15-20

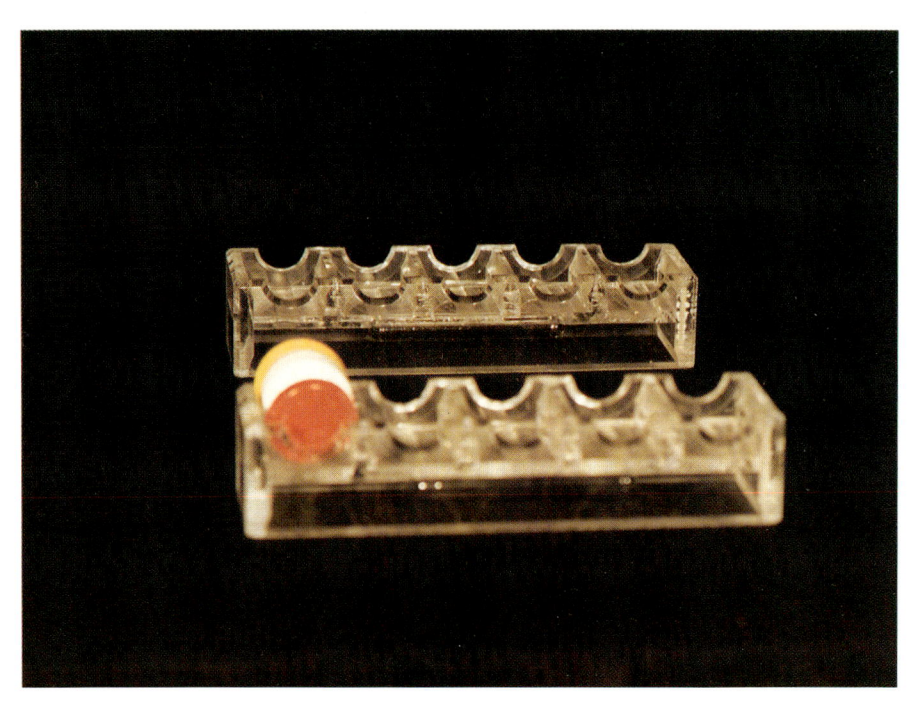

Lucite chip holder. 1950s-1960s. $40-45

Lucite chip holder with plastic chips. 1950s. $65-85

Boudoir Items

Light blue dresser set trimmed in rhinestones and plastic flowers. "Menda Co., Pasadena, Ca.," 1950s. $18-22

Light pink hair receiver trimmed in plastic flowers and rhinestones. "Menda Co., Pasadena, Ca.," 1950s. $20-25

Light blue powder box trimmed in rhinestones and plastic flowers. "Menda Co., Pasadena, Ca.," 1950s. $10-14

Gold plastic jewelry box. "Japan," 1950s. $20-25

Swivel jewelry box in green plastic. 1970s. $20-25

Plastic trinket box with mirror. 1970s. $9-12

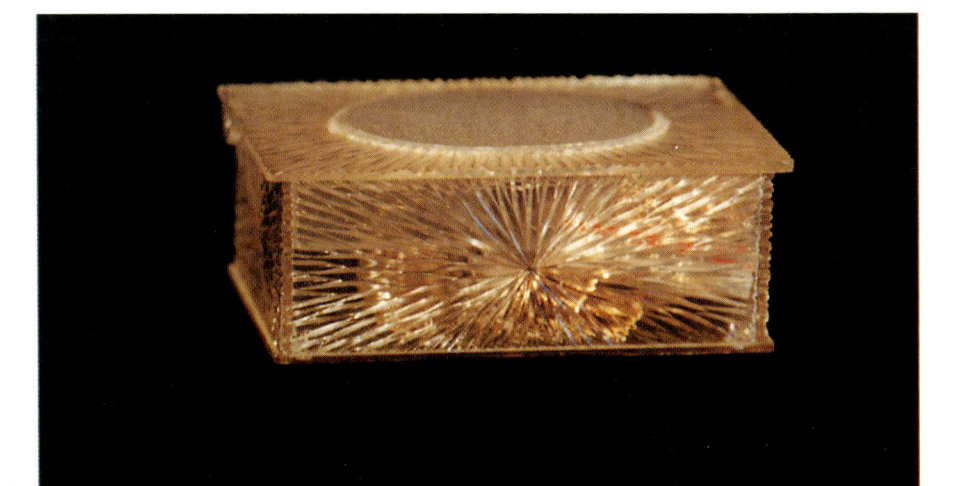

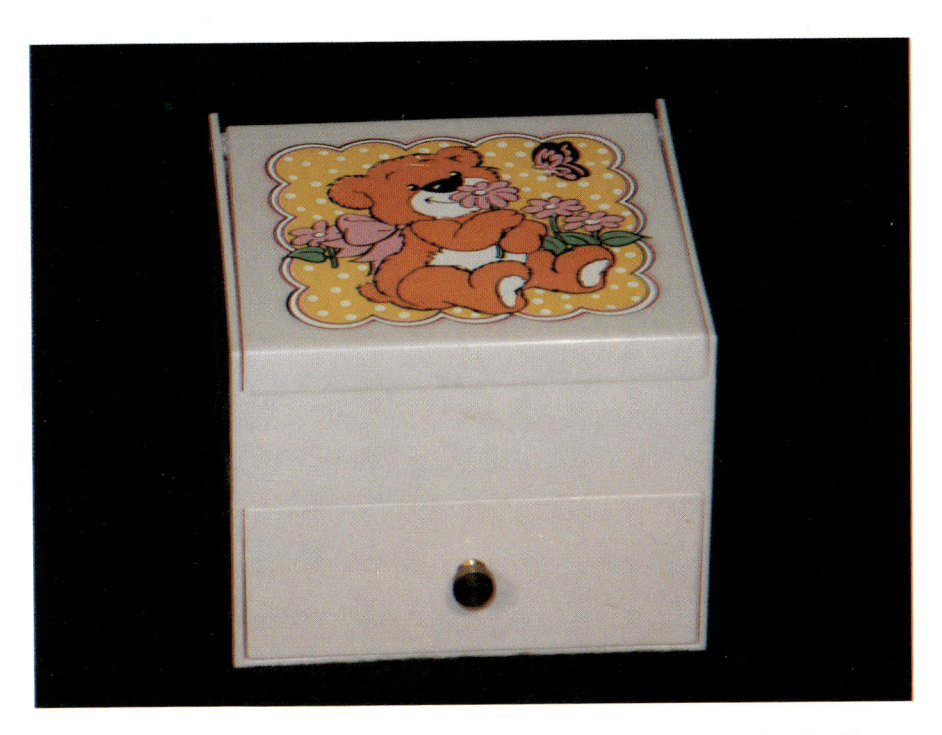

Child's jewelry box decorated with bear on top. Mirror inside top lid and pull out drawer. "Made in Hong Kong," 1970s. $10-15

Lucite violin jewelry box. 1950s. $20-30

Gold colored plastic box with lid. "Creative Containers Corporation," 1950s. $15-20

Radio-styled musical trinket box with plastic knobs. 1950s. $20-25

Lucite and gold glitter tissue box. "Blisscraft of Hollywood," 1950s. $20-25

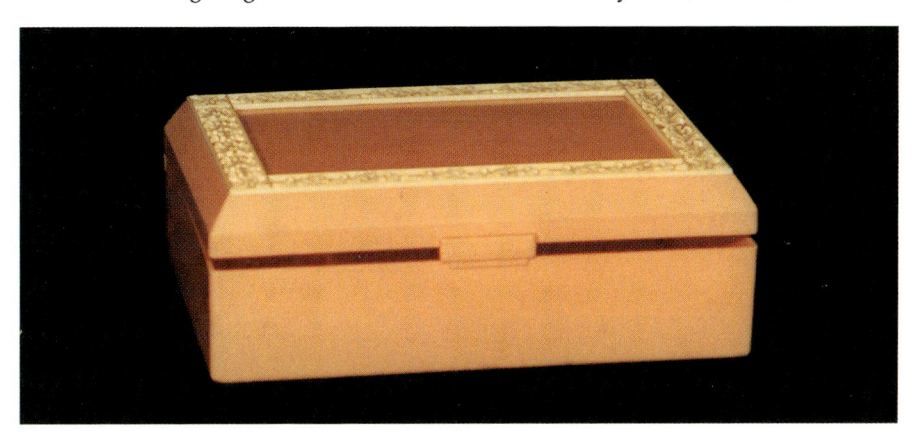

Pink tissue box with white trim. 1960s. $8-10

Thin shell with plastic frame tissue box. 1970s. $18-22

Tortoise shell compact with mirror inside lid. Made in Hong Kong. 1960s. $20-25

Clear Lucite brush set. 7" and 8". 1950s. $18-22 set

Plastic tooth brush holder from "Schwartz Bros. Plastics." 1970s. $7-10

Plastic soap dish with carved plastic flower on front. "Schwartz Bros. Plastics," 1970s. $8-10

Glove Tree glove dryers. 1950s. $20-30

Bakelite button hook. 1920s. $45-50

"Lusterware" all purpose clothes hangers. 1960s. $5-7 each

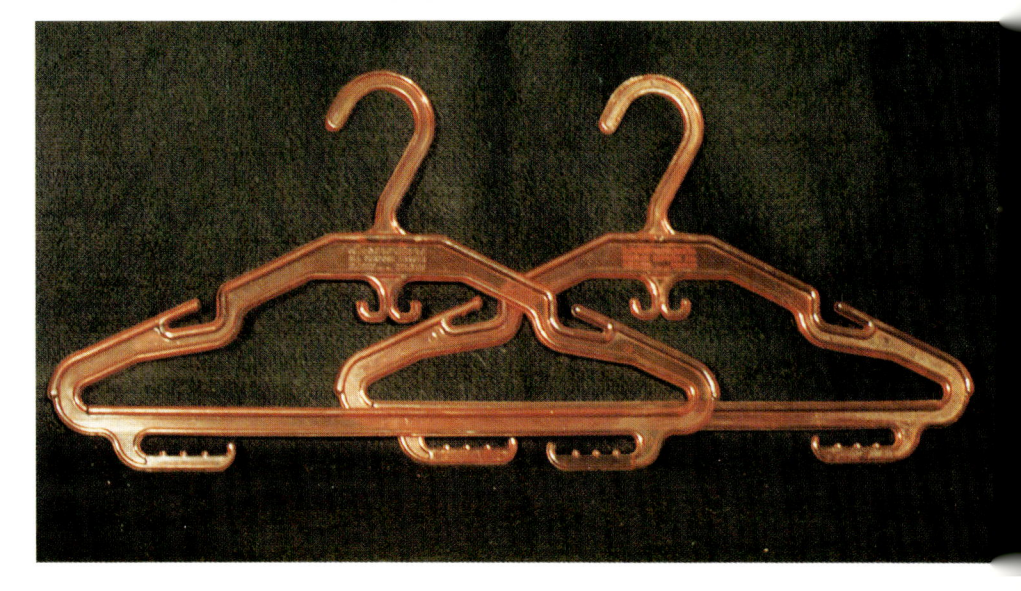

Wall Hangings

Plastic woven-look wall plaques. "Burwood Products Co." 1970s. $10-16

Plastic letter holder with flower design. 1970s. $5-8

"Reliance" plastic bead plant holder. 1960s. $10-15

Green and white plastic vase with pierced design on outer layer. "Macrame Line, E.O. Brady Co.," 1970s. $7-10

Red plastic wallpocket tea cup and saucer. "U.S.A.". 1950s. $12-15

Plant holder with red plastic beads. 1960s. $12-15

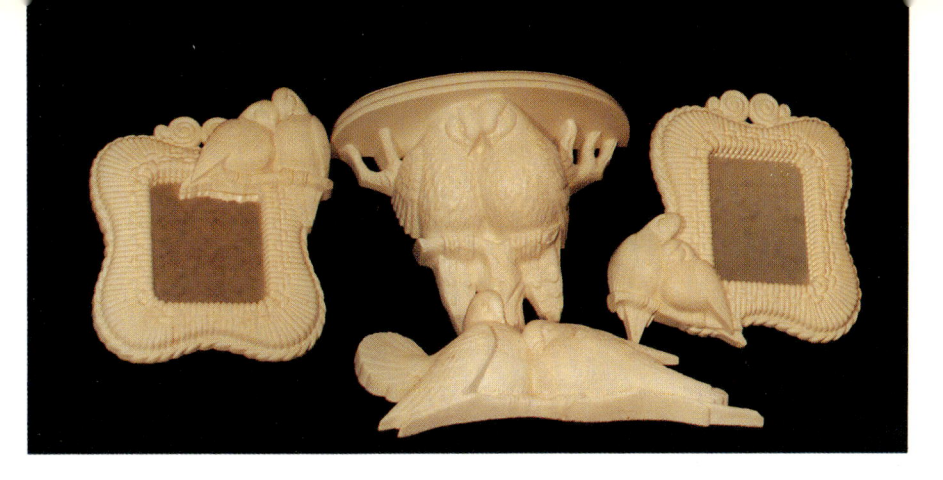

Burwood Products Co.'s plastic wall shelf with mirrors and doves. 1970s. $35-45

Plastic wall sconce set, by Universal Statuary Corporation, Chicago. 1962. $35-45

Plastic wall pockets. "Syroco Inc. U.S.A," 1970s. $20-25

Plastic wall plaques with applied flying Mallards, also plastic. 1950s. $10-15 pair

Plastic wall plaque with floral design. 1970s. $20-25

Pair of plastic, gold sprayed, floral design wall plaques. 1970s. $15-18

Plastic set of three diamond-shaped wall plaques. "Dart, Indiana, Made in U.S.A". 1970s. $15-20 pair

Pair of plastic wall plaques with Roman designs. 1970s. $12-16 pair

Gold washed plastic Roman ruins wall plaques. "Homco, made in U.S.A". 1970s. $15-20 pair

Electrifying Plastic

Tulip-form electric wall sconce. Plastic, 1960s-70s. $40-50

Tulip-form plastic wall light. 1960s-70s. $40-50

Hanging light. Lucite with spherical chrome globes. 1950s-60s. $225-250

Metal Spanish-style hanging light with plastic liner. 24". 1950s-60s. $65-75

Yellow woven-look plastic hanging ceiling light. 1970s. $30-40

Hanging electric light with plastic grapes and leaves. 21" high. 1950s. $225-280

"Seth Thomas" electric clock. 1950s. $18-22

Advertising Items

"Joe Camel" battery operated clock. 1990s. 23". $80-100

Garfield battery operated clock. "Sunbeam," 1978. 17". $60-80

Garfield battery operated alarm clock. "Sunbeam," 1980s. 18". $75-100

107

Marlboro wall clock, battery operated. 1991. $40-50

Orange plastic one foot ruler. 1960s. $10-15

Plastic Jimmy Carter peanut bank. $30-40

"Planters Peanut" plastic cup. $20-25

Plastic "Mr. Peanut" standing figure. 1991. $10-15

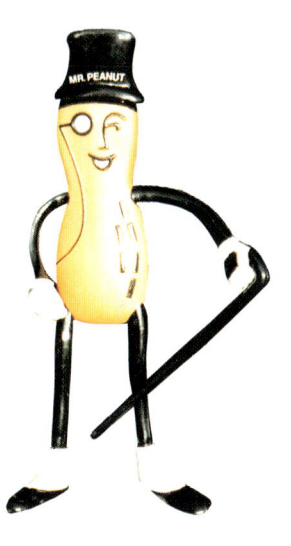

Swivelling comb display case. "Ajax Combs". 1940s. 18". $40-50

Dancing "Diet Pepsi" can. Moves to the sound of music. 1990s. $15-20

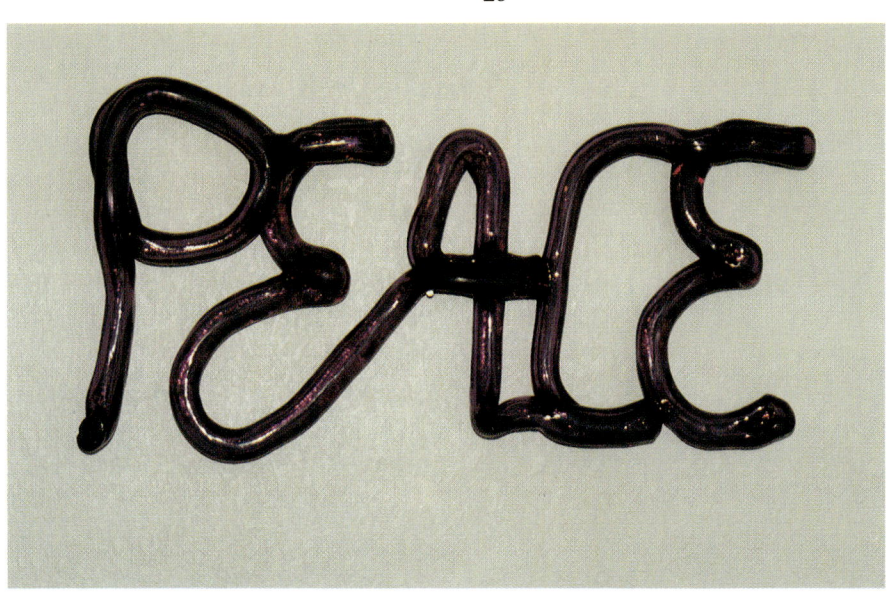

Blown plastic peace sign. 15". 1950s. $30-40

"Campbell Soup" plastic play set. Marked "Prossed Plastic Co." $20-30

"Campbell Soup" insulated cup. 1960s. $20-30

"Campbell Soup" child's plastic cup and plate. $15-20 set.

Banks & Toys

Plastic daisy design piggy bank. "Universal Statuary Corp., 1976". 19" x 12.5". $80-100

"Robie The Robot" mechanical bank. 1960s. (Made again in 1990s.) Put money on hand and he flips it into his mouth and licks his lips. $40-45

Money-Hungry Bank, "Put your money where your mouth is." The money is placed on slot, which is pushed down. Then the money goes into mouth and the teeth chatter. 1975. $30-40

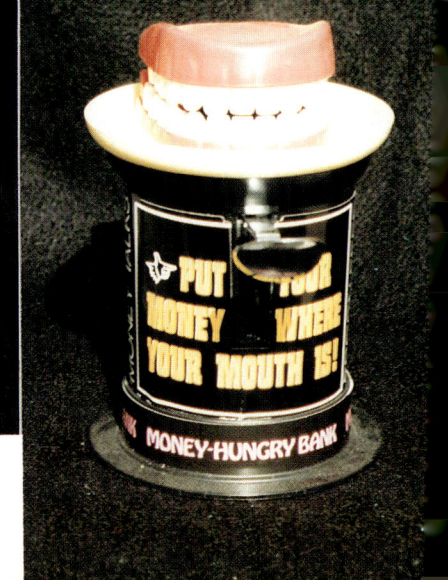

"Tea Time" plastic music box with kitchen scene. 1960s. $15-20

Musical windmill. "Hong Kong," 1960s-70s. $20-30

Musical chimney with Santa on front. "Yuletide Enterprises," 1970s. $20-30

Angel music box. 1960s. $18-22

Celluloid doll carnival prize. 1950s. $35-45

Calypso Queen plastic doll. 11". "Nassau," 1950s. $30-40

Plastic "Hi Tech" yo-yo. 1960s. $20-25

"Oscar Mayer" plastic whistles. $6-8 each

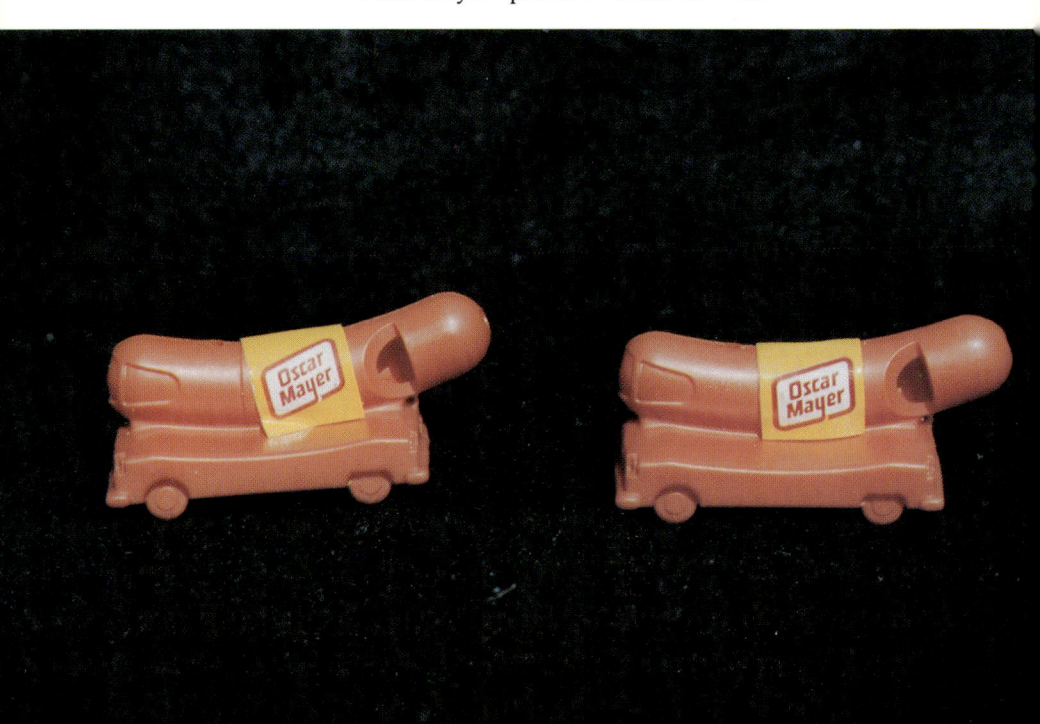

Miscellaneous Items

Clear Lucite shoe in original box. "Coty". 1940s. $30-35

Plastic party picks and holder. 1970s. $4-7

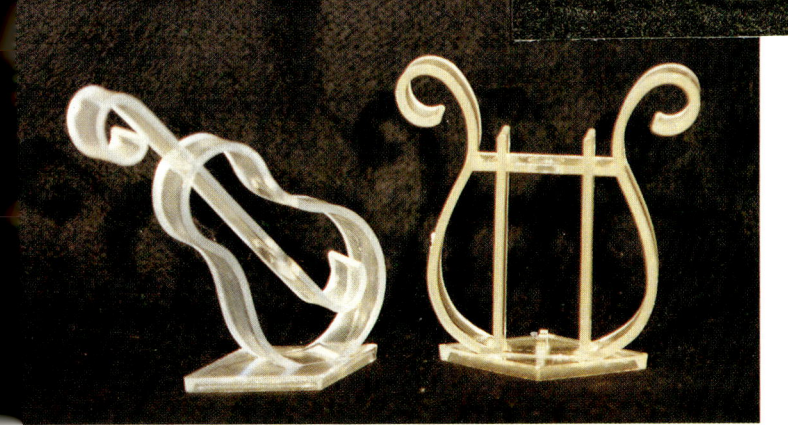

Telephone candy container. 1950s. $20-25.

Lucite harp and violin candle holders. "Made in Hong Kong," 1970s. $10-15 each

117

Candle holders and candles, clear plastic with foil design. 1960s. $15-20 set

Clear plastic candle holder with green tapers. 1960s. $20-25 set

Pair of plastic sitting duck decoys. "Fairfax, Feather Lite, J.S.McGuire," 1954. $15-20 each

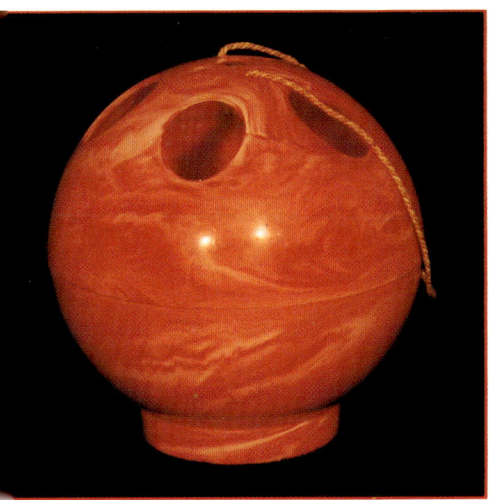

Swivel fern holder. Marble-like plastic, 1970s. $5-8

Musical phone message keeper. "Japan," 1950s. $18-25.

Plastic wall clock. "Spartus," 1970s. $15-20

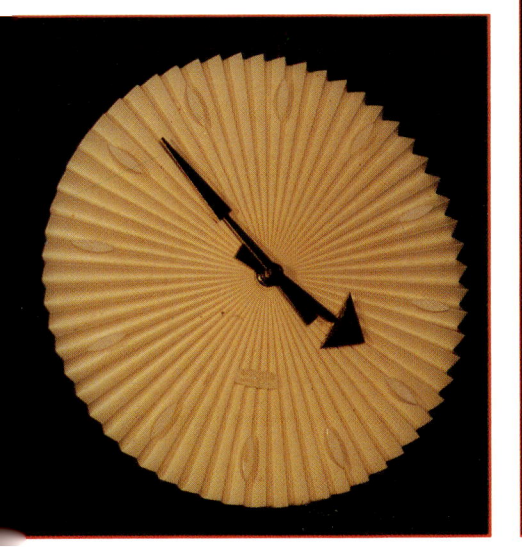

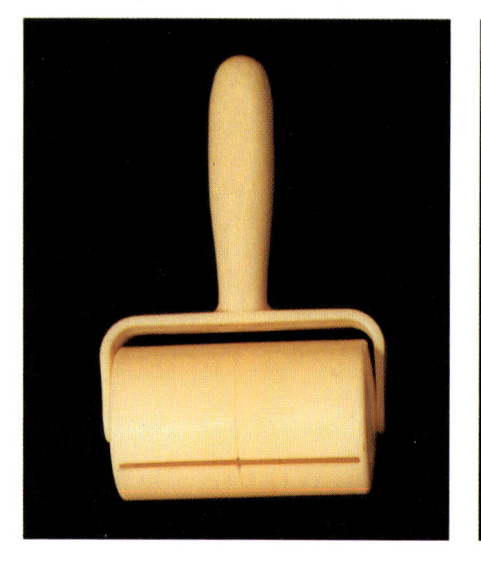

Lint remover. 1960s. $7-10

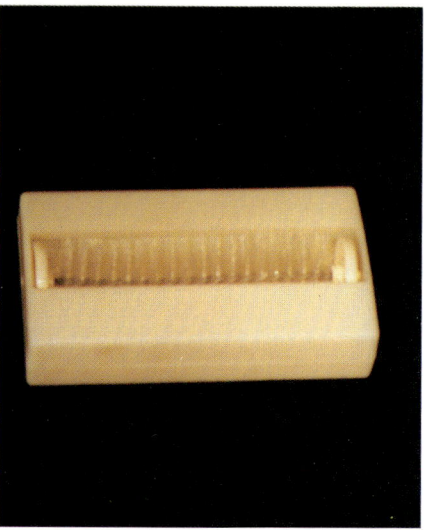

Crumb brush. 1960s. $6-8

"Suntarce" plastic tape dispenser. 1960s. $10-15

Refrigerator deodorizer with decorated with lemons. All plastic, 1960s. $6-10

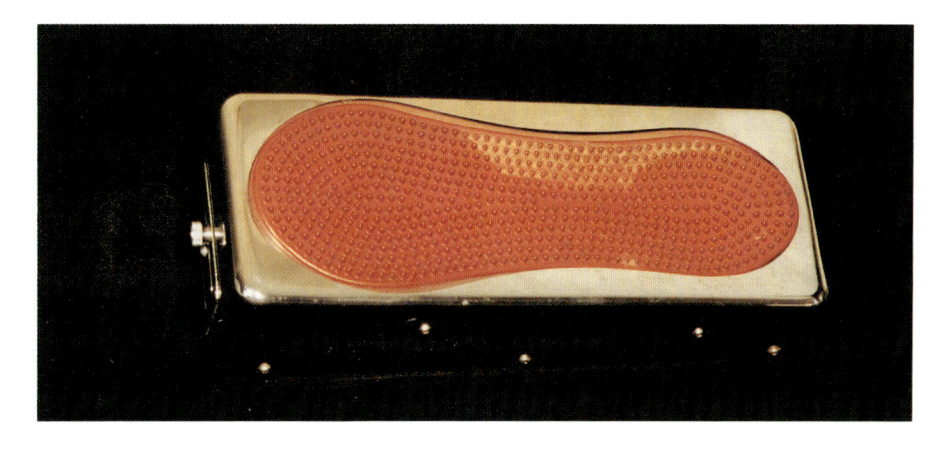

Executive foot massager by "Swank". Chrome battery operated with soft foot pad insert. 1950s. $35-45

Cotton pin cushion with plastic wrist attachment. 1960s. $6-8

"Singer" sewing machine parts cases. 1960s. $15-20

"Jack LeLanne Twist-Away Exerciser" Plastic, 1960s. $35-45

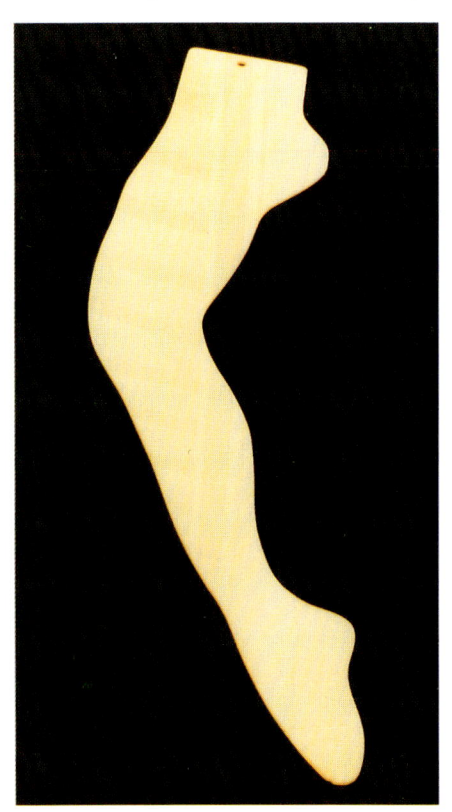

Hosiery form for store display. Plastic, 36" long. 1950s. $40-50

"Venus Di Milo," amber colored resin. 1960s. $65-70

Plastic portable bar with cup inserts and 2 shelves. Cube design. 1960s. $60-75

Pair of Lucite frame chairs with velvet seats and chrome pedestal base. 1960s. No price available.

Personal Plastics

Purses

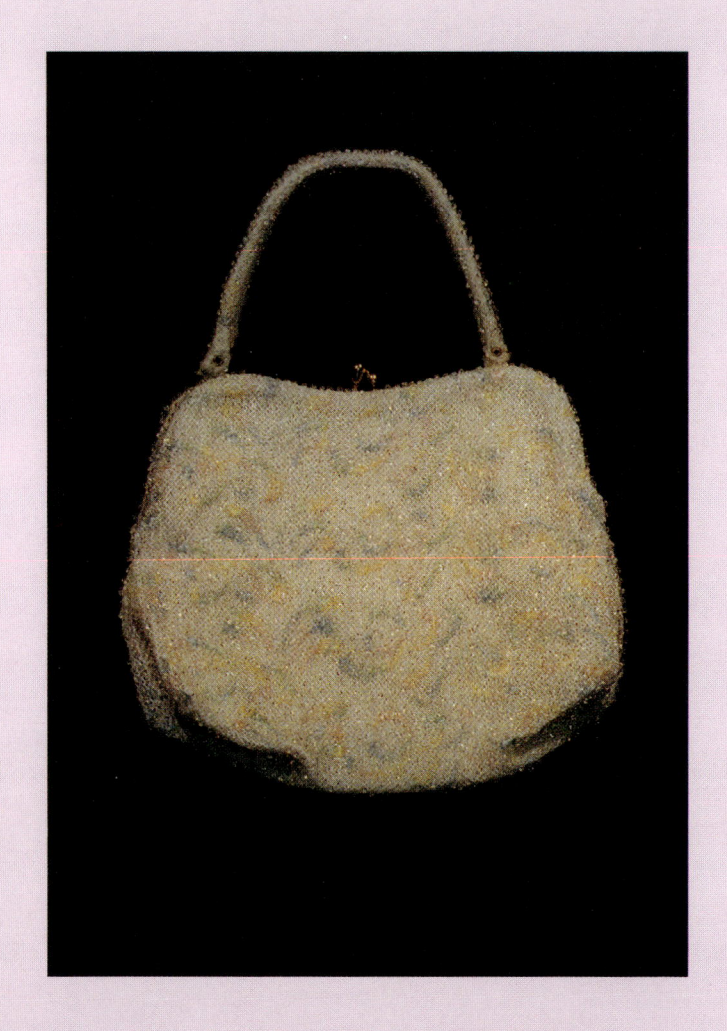

Plastic beaded purse and handle. 1960s. $15-20

Plastic Lucite pill box purse with Lucite handle and gold wash clasp. 1950s. $50-60

Orange Lucite purse with handles. 1940s-50s. $70-80

Lucite pill box purse. 1950s. $40-50

Beaded plastic weave handbag. 1950s. $20-25

Plastic beaded purse with metal handle and frame. 1960s. $15-20

Beaded plastic handbag with rhinestone inserts. 1960s. $20-25

Cord purse with Lucite handle and clasp. 1960s. $40-50

Lucite purse with gold wash clasp and decoration. 1940s-50s. $60-75

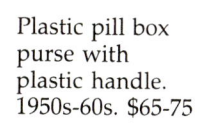

Plastic pill box purse with plastic handle. 1950s-60s. $65-75

Gold weave box purse with Lucite handle. 1960s. $35-45

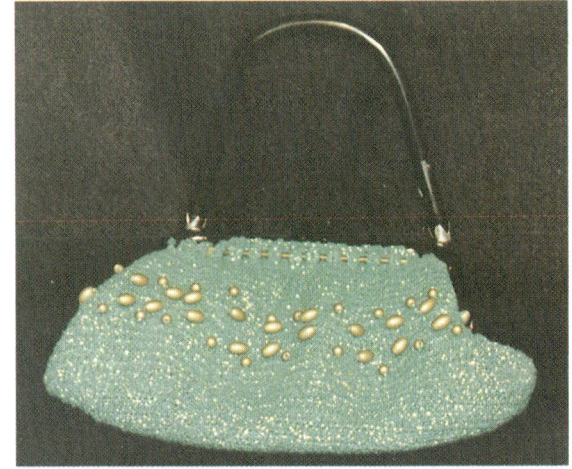

Plastic beaded handbag with plastic handle. 1950s-60s. $18-25

Plastic see-through purse with floral design and plastic handle. 1950s-60s. $15-20

Bracelets

Multi-colored swirled plastic bracelets. 1970s. $3-4 each

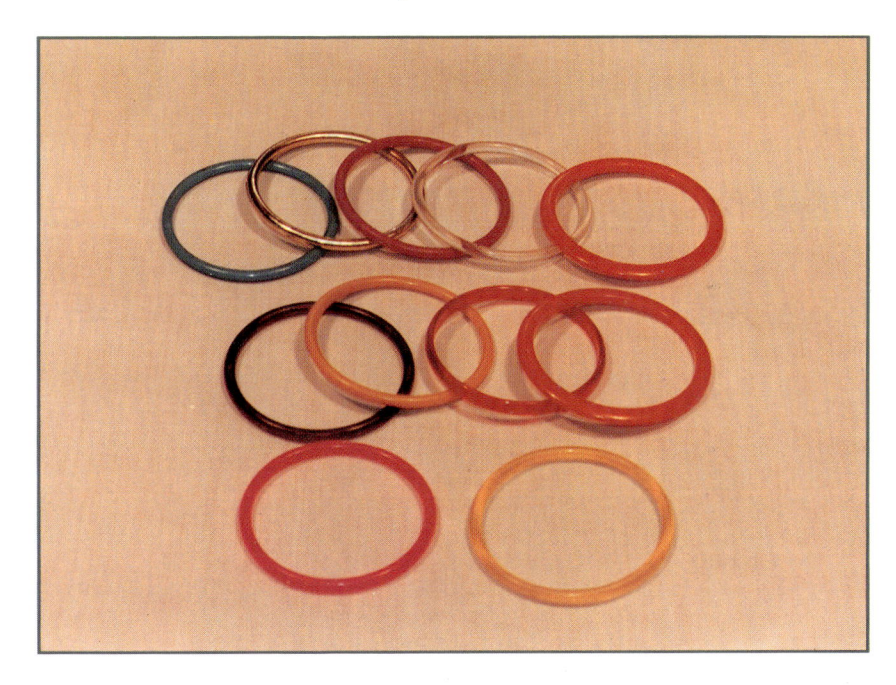

Set of eleven multi-colored plastic bracelets. 1970s. $2-3 each

Inlaid blue and black plastic bracelet. 1960s. $25-30

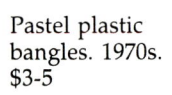

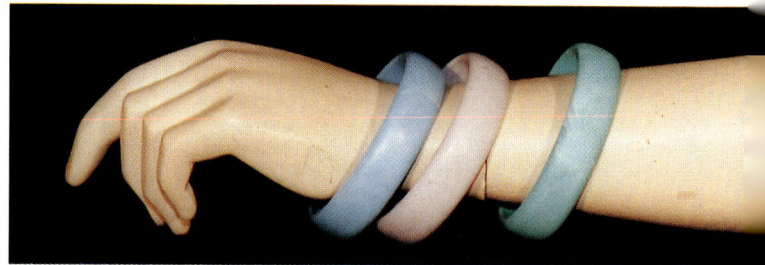

Pastel plastic bangles. 1970s. $3-5

Carved Bakelite wide band bracelet. 1960s-70s. $30-40

Yellow Bakelite bracelet. 1950s. $40-50

Multi-colored round bracelets. Plastic in assorted sizes, 1960s. $4-5 each

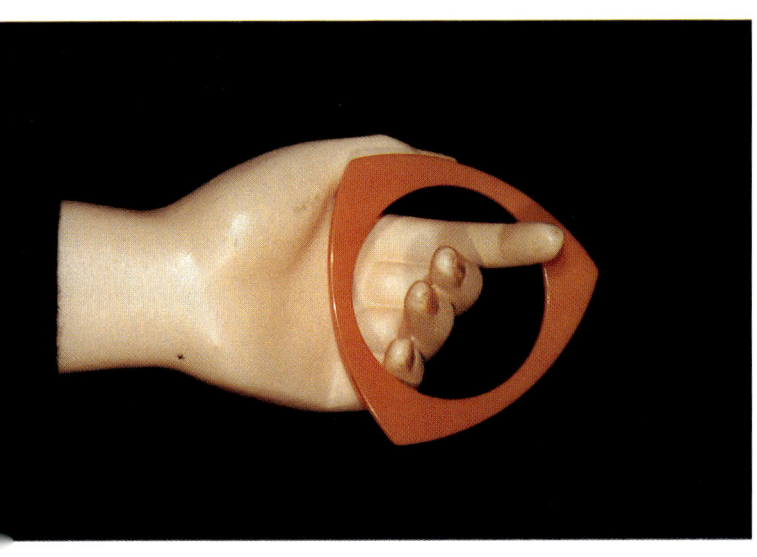

Triangular Bakelite bracelet. 1960s. $15-20

131

Gold, ivory, and orange plastic stretch bracelet. 1960s. $20-25

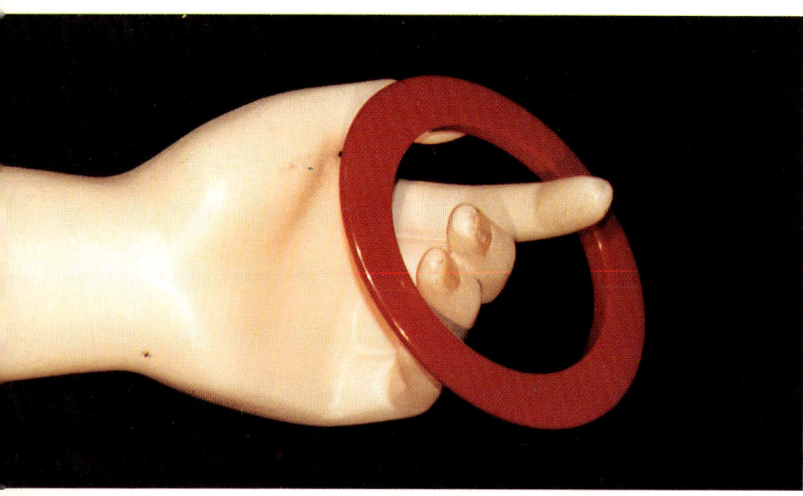

Blood red oval plastic bangle. 1970s. $8-10

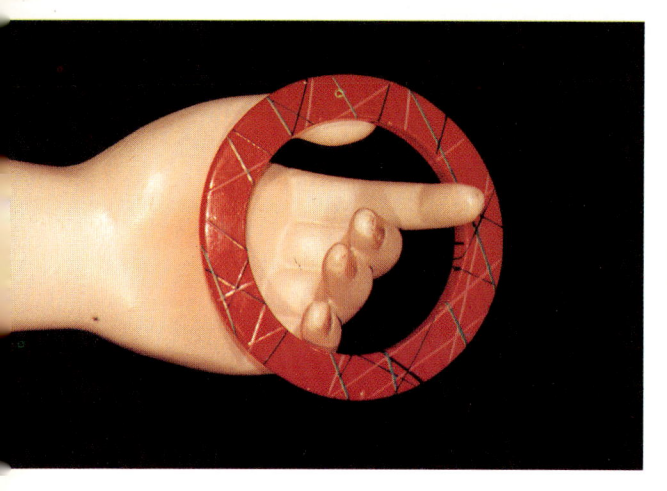

Red plastic bangle with string design. 1970s. $5-8

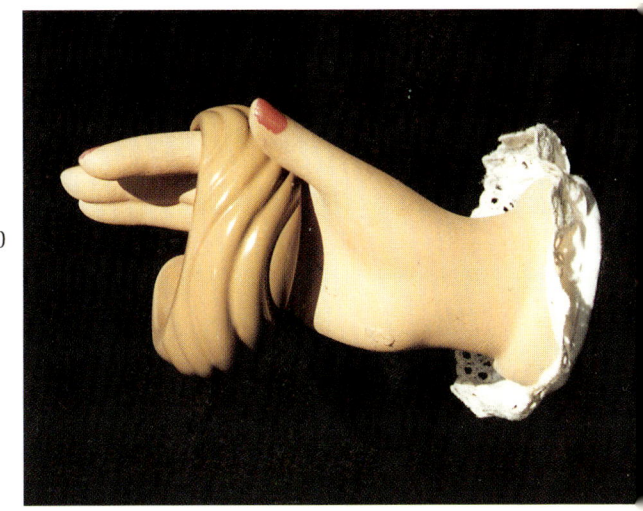

Plastic swirl bracelet. 1960s. $15-20

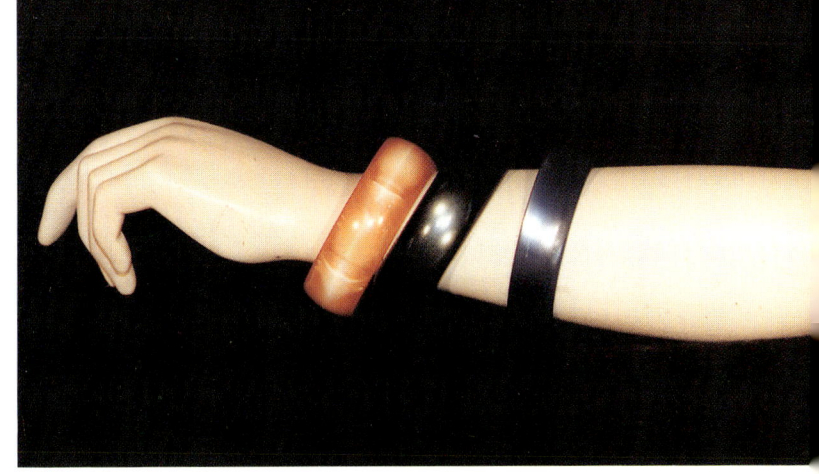

Set of three multi-colored marbleized plastic bracelets. 1950s. $35-45 set

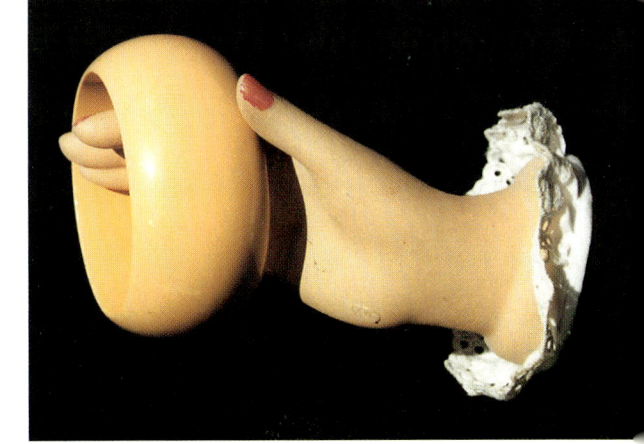

Wide band plastic bracelet. 1960s. $18-22

Round and diamond shaped plastic dangles on chain bracelet. 1950s. $8-12

Tear drop design plastic bead bracelet. 1950s. $10-15

Swirled multi-colored plastic bracelet on elastic string. 1960s. $15-20

Necklaces

Metal necklace and bracelet with inlaid plastic. 1950s. $20-30 set

Designer necklace, plastic pendent on gold wash chain. 1960s-70s. $10-15

Red plastic "Pop" beads. 1950s. $15-20

Plastic disc beads on waxed string with sterling silver clasp. 1960s. $25-30

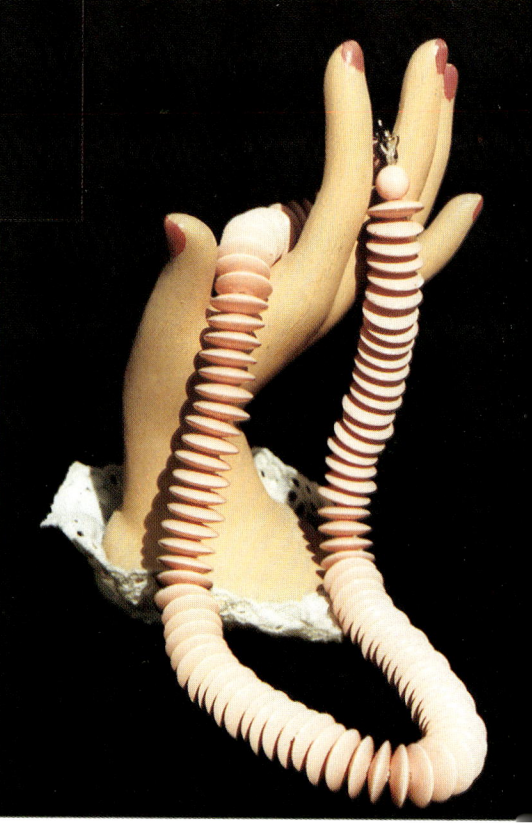

Earrings

Gold tone sphere-shaped pierced earrings with multi-colored stones and gold wash drop chains. 1960s. $15-20

Plastic oval clip-on earrings with embedded rhinestones. 1950s. $15-18

Two pairs of clip-on plastic beaded earrings with large pearlized plastic stone in center. 1950s-60s. $5-8 pair

Clip-on earrings with pearlized plastic beads in a rosette pattern. 1960s. $4-6

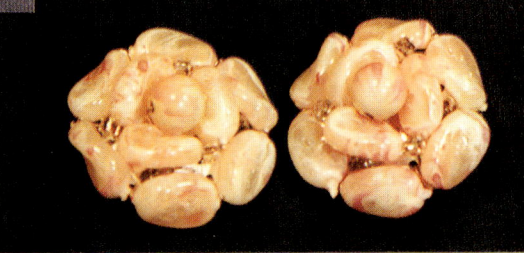

Screw-on plastic earrings with pearl in center of flower. 1950s. $5-7

Six pairs of plastic beaded clip on earrings. 1950s. $3-5 pair

Clip-on plastic flower earrings with rhinestone set in center. 1960s. $8-10

Leaf-shaped screw-on earrings with rhinestone center. 1950s. $10-15

Plastic beads strung with gold wire on metal base. Large round pearlized plastic center. 1960s. $8-10

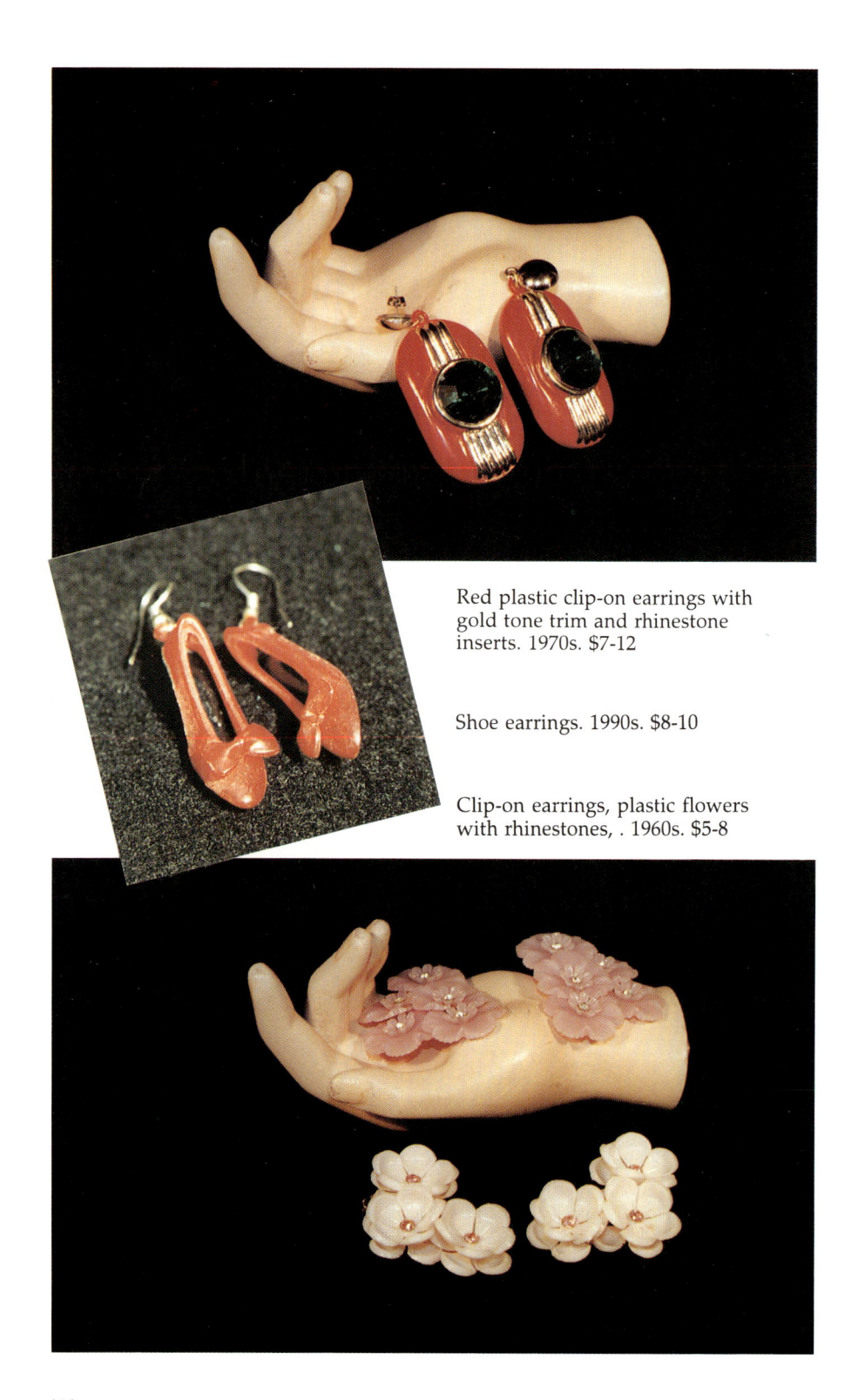

Red plastic clip-on earrings with gold tone trim and rhinestone inserts. 1970s. $7-12

Shoe earrings. 1990s. $8-10

Clip-on earrings, plastic flowers with rhinestones, . 1960s. $5-8

Floral clip-on earrings, plastic and rhinestone. 1950s. $7-10

Screw-on plastic earrings with beads set in center of petals. 1960s. $5-8

Plastic and pearl beads on clip-on earrings. 1950s. $7-10

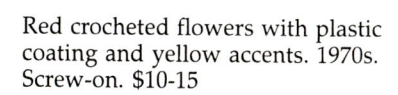

Red crocheted flowers with plastic coating and yellow accents. 1970s. Screw-on. $10-15

Red teardrop cluster, pierced dangle earrings. 1960s. $12-18

Clip-on earrings with plastic and rhinestone centers. 1950s. $8-10

Clip-on earrings, gold tone rope with large red plastic centers. 1970s. "Made in Italy." $5-8

Screw-on plastic tear drop earrings. 1960s. $10-14

Belts

Multi-colored woven wrap belt, with plastic beads. 1970s. $15-20

Tortoise shell plastic large disc belt with brass chain drop. 1960s-70s. $15-20

Gold-toned metal chain belt with ivory colored plastic stones. 1970s. $7-10

Two-tone plastic disc belt. 1960s. $15-20

Tortoise shell plastic belt, rectangle segments with dangling chain. 1960s. $15-20

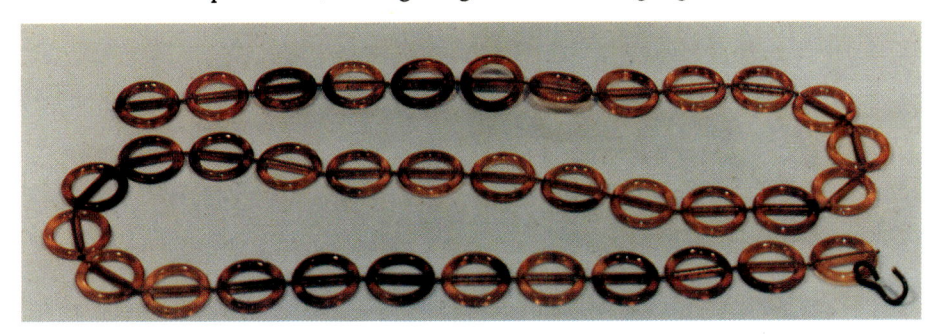

Round tortoise shell plastic belt with hook fastener. 1960s. $12-15

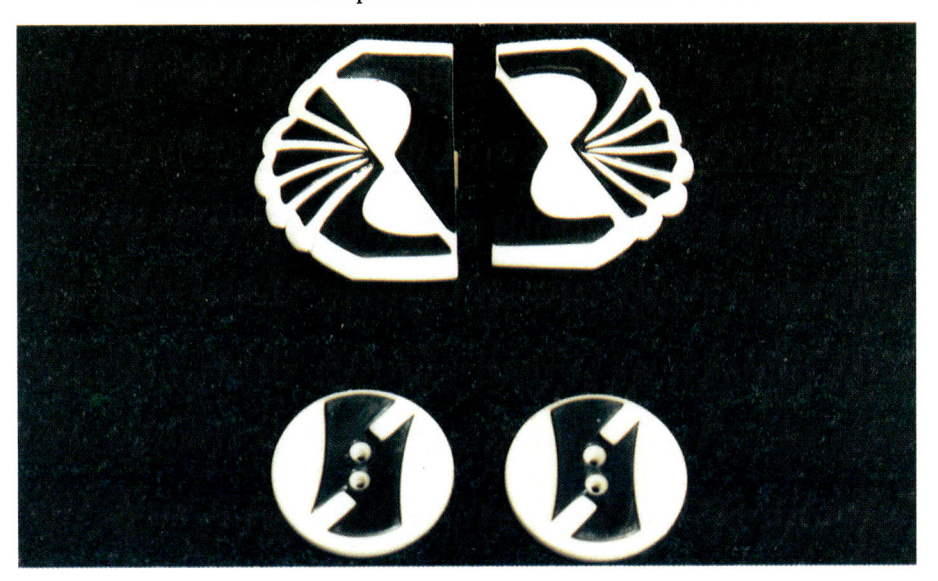

Matching carved plastic buckle and buttons. 1930s. $10-14

Pins

Plastic Toucan pin with rhinestone eyes and chest. 1950s. $40-45

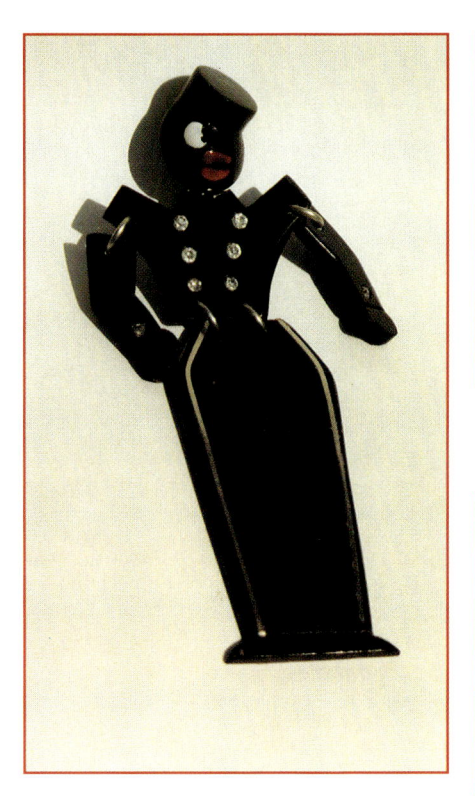

Bakelite black porter pin. 1940s. $250-275

Beaded intertwined circle pin. 1960s. $8-10

Lace-like broach, plastic with rhinestones in the centers of flowers. 1960s. $20-25

Pin of elephant riding a bike. "Avon". 1975. $7-10

Hat broach. 1950s. $30-40

Parrot pin, plastic with painted eyes. 1980s. $20-25

Hair Clips

Plastic alligator barrette. 1970s. $6-8

Plastic hair combs. "Model Repose France," 1970s. #2-3 each

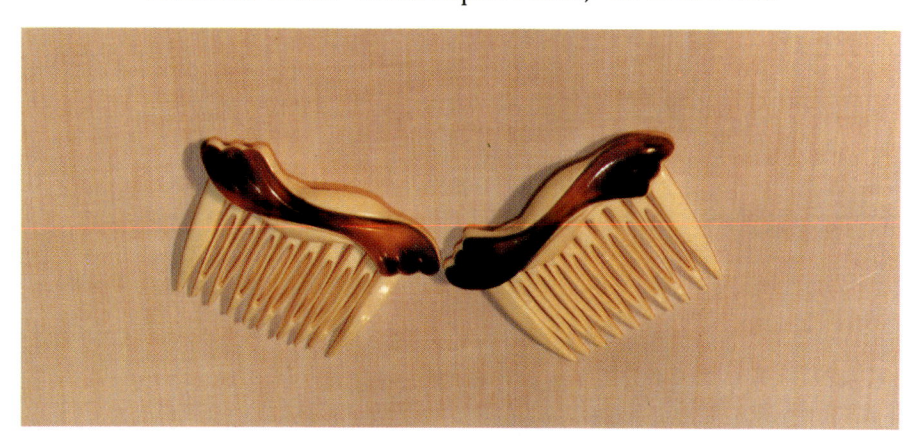

Tortoise shell hair combs. "France," 1960s. $10-15 pair

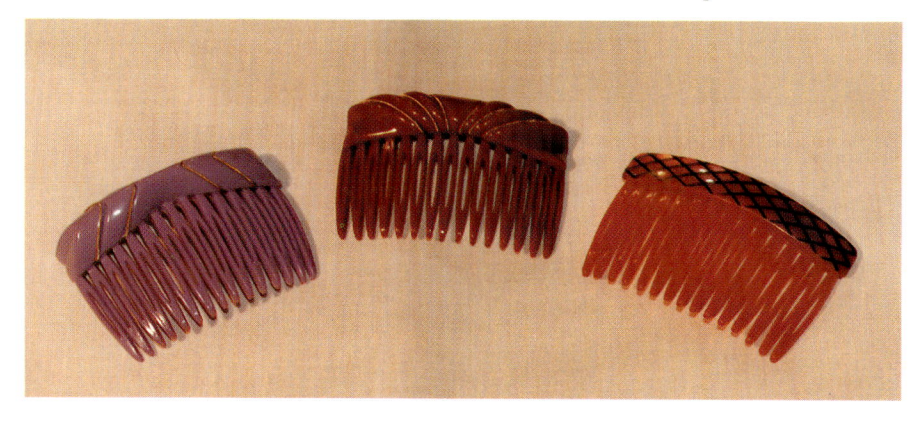

Plastic hair combs. 1980s. $2-3 each

Silver frame hair comb with plastic stone and teeth. $5-7

Plastic bow barrette with pearl drops on metal frame. "Korea," 1970s. $5-7

Bow-shaped plastic hair comb. 1970s. $4-6

Red and black barrette. 1980s. $5-7

Barrette with mod design. "Philips," 1970s. $6-8

Sunglasses

"Cat eye" sun glasses with gold trim. 1950s. $15-20

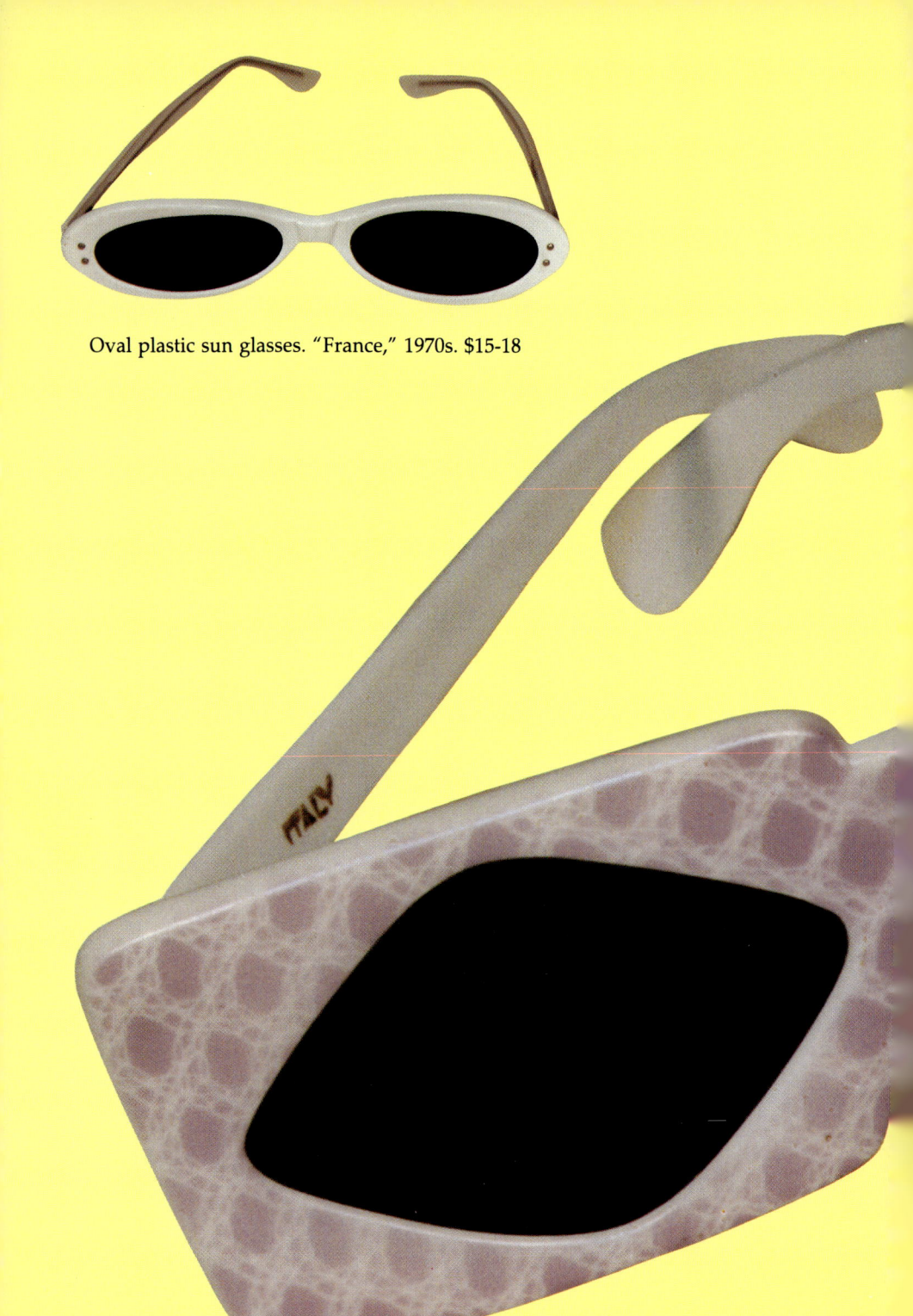

Oval plastic sun glasses. "France," 1970s. $15-18

Square frame sun glasses with diamond shaped lenses. 1960s. $25-35

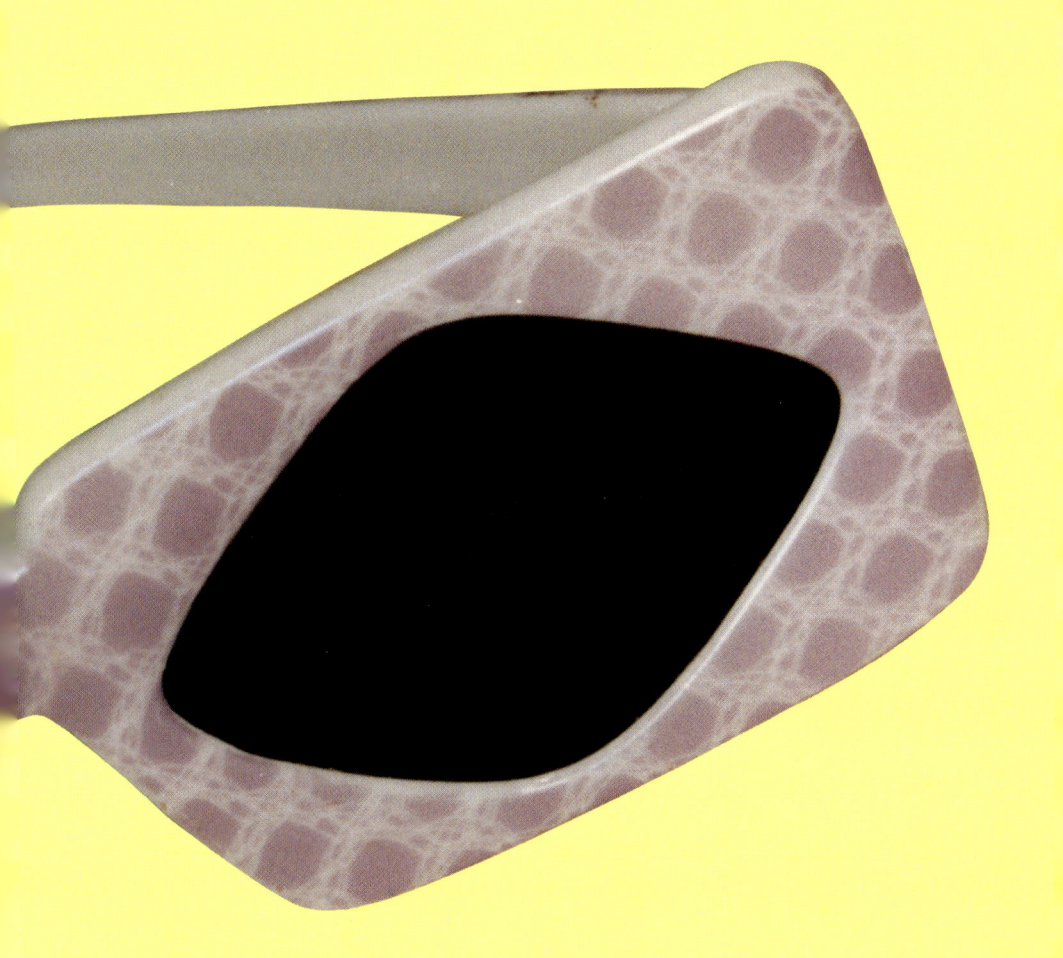

Square-framed sun glasses. "Italy," 1960s. $12-17

Heart-shaped plastic sun glasses with heart ear bobs. 1970s. $40-45

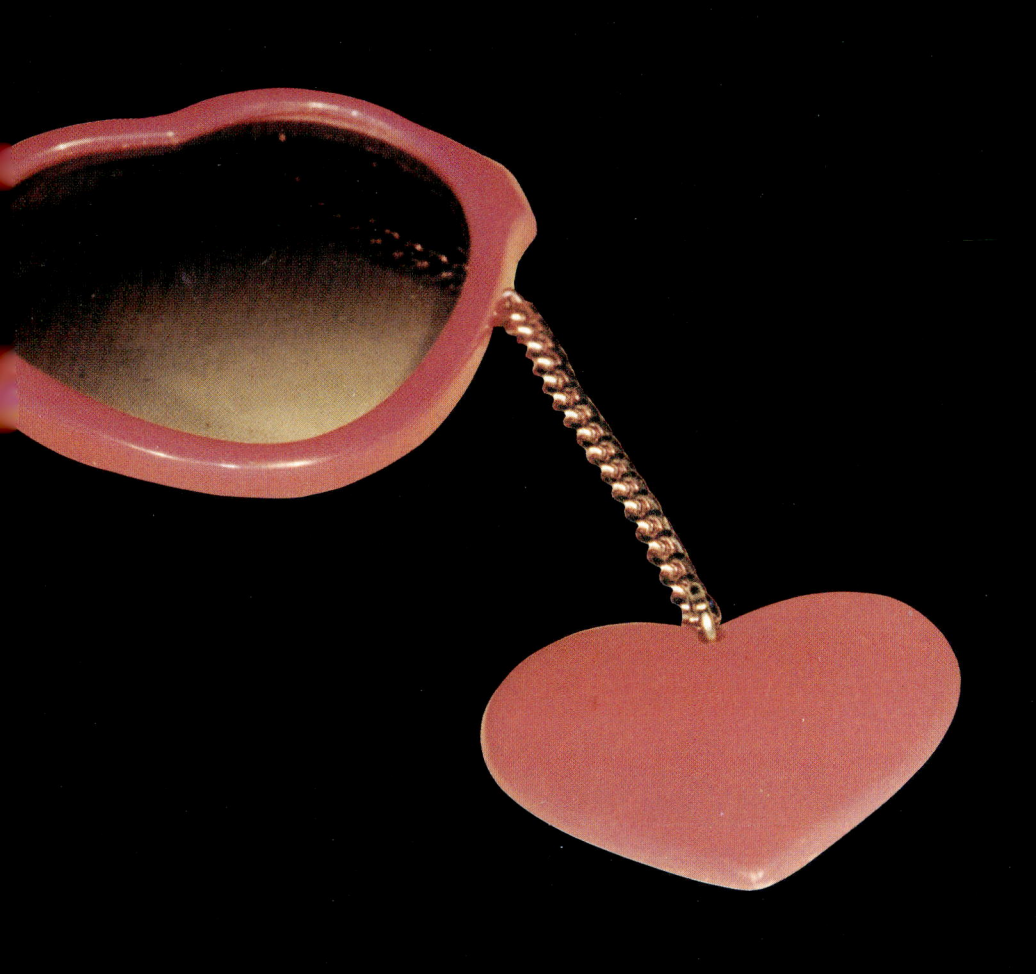

Double frame sun glasses. 1960s. $18-22

Clear purple oval plastic sun glasses. "Bachman-Italy," 1970s. $8-12

Translucent green oval sun glasses. "Bachman-Italy," 1960s. $10-15

Plastic checkered sun glasses. "Italy," 1950s. $20-25